Atoms, Electrons, and Change

ATOMS, ELECTRONS, AND CHANGE

P. W. Atkins

SCIENTIFIC AMERICAN LIBRARY

A division of HPHLP
New York

Library of Congress Cataloging-in-Publication Data

Atkins, P. W. (Peter William), 1940–
  Atoms, electrons, and change / P. W. Atkins.
    p. cm.
  Includes bibliographical references and index.
  ISBN 0-7167-5028-7
  1. Chemical reactions. 2. Atomic theory. I. Title.
QD461.A84 1990
541.3′9—dc20   91-12336
                                     CIP

ISSN 1040-3213

Printed in the United States of America

Scientific American Library
A division of HPHLP
New York

Distributed by W. H. Freeman and Company
41 Madison Avenue, New York, New York 10010
20 Beaumont Avenue, Oxford OX1 2NQ, England

1 2 3 4 5 6 7 8 9 0   KP   9 9 8 7 6 5 4 3 2 1

This book is number 36 of a series.

CONTENTS

PREFACE

Nature is intricate, but local patches of nature, at least, are not beyond understanding. In particular, all the transformations of matter, including the countless changes that surround us and take place within us as we grow and participate in the activities of life, can be comprehended in terms of the rearrangements of atoms. Such transformations of matter, which may lead to the unfurling of a leaf in one location, the incandescence of a flame in another, and the emergence of a flavor (or even the formation of a thought) elsewhere, are the consequences of chemical reactions; indeed, such processes *are* chemical reactions. It is through an understanding of the events that occur when a reaction takes place that we can edge a little closer to understanding the almost overwhelming complexity of this extraordinary, complex, yet at root simple world. Of course, we may need to observe from a different vantage point if we are to see the collective outcome of the changes that occur when atoms move, for we cannot yet trace the formation of an opinion to the motion of an atom. Yet by arming ourselves with an understanding of the atomic basis of chemical change, we are at least acquiring a depth of comprehension that adds to our appreciation of nature.

I aim to show how to comprehend the changes brought about by chemical reactions by exploring how the electronic structure of an atom determines its chemical destiny. As we travel in our imagination through the wall of a flask and put our eye ever more closely to the events occurring within, we shall see atoms exchanging partners, molecules forming progressively more sophisticated structures, atom attacking atom, groups of atoms being squeezed out of a molecule, and chains of atoms writhing and twisting into rings. By understanding how an atom can be moved from one location to another, we come to see the atomic texture of the tapestry of change, and it will be possible, in a general sense if not in great detail, to appreciate the world around us on a finer scale than mere appearance.

However, because a description of *all* chemical reactions would be a description of the whole of chemistry, and bloat volumes, you should not expect to find here a description of all possible chemical transformations. I have had to be severely selective. I have chosen to present a series of linked accounts that describe different aspects of chemical reactions, and through them to show a little of how chemists think about those different aspects—including the classification of reactions, their driving force, their rates, and their mechanism. In particular, I have taken carbon as the target of my attention, for as well as being crucially important to life, it displays a great many facets of chemical reactions.

For the vehicle of my discourse, I use Michael Faraday's *Chemical history of a candle*. Faraday was among the most successful exposers of chemical reactions to the previously unseeing, and in his famous series of lectures he took the mundane and showed that it was a microcosm of chemical change. Moreover, Faraday lived a century and a half ago, and it is of interest—and provides a framework for the discussion—to see how far we have moved beyond the pale of his understanding.

Faraday's views were grounded in early nineteenth-century chemistry, when the subject was barely formed and many of its central concepts were still awaiting discovery. He knew enough about the everyday appearances of matter to make an engaging story, and he was the arch-discerner of appearances and the epitome of common sense. Yet now we have the advantage of knowing that unguided common sense may be misleading, for when our discourse touches atoms, we have to use concepts that were entirely alien to Faraday's world: specifically, we have to use quantum mechanics. We shall see that the more closely we scrutinize Faraday's candle with a modern eye, the more classical concepts melt like the candle's wax and the elusive flickering shadows of quantum mechanics become essential to elucidation.

Such is the range of concepts a modern candle can introduce, that, unlike Faraday, I could not have written this book without a great deal of help. I am very grateful to Robert Crabtree at Yale and Maitland Jones at Princeton, who both gave technical help and valuable advice by reminding me what and whom I should have in mind. I am grateful to Gary Carlson, whose imaginative grasp of chemistry is an inspiration and who first suggested that I should stretch myself on this particular rack. The wheel of the rack was turned by Susan Moran, that most professional of editors, who sent me back to the dungeon more times than I care to remember, but always will. I also thank Travis Amos, from whom I have learned so much before, and who once again brought his unparalleled visual sensitivity to bear on the selection of photographs. I am very grateful to Diane Maass, who guided the book through the intricacies of its production with care and thoughtful understanding.

P.W.A.
Oxford, February 1991

Atoms, Electrons, and Change

UNLEASHING THE FLAME | 1

The destructive impact of a forest fire, while transiently tragic, is a source of carbon dioxide that in due course will be converted, once again, into vegetation. Faraday's candle is a conflagration on a smaller scale, and it too released carbon atoms (as carbon dioxide) into the atmosphere but in a less dramatic style. In these pages, we shall follow the chemical history of these atoms.

I propose to bring before you, in the course of these lectures, the Chemical History of a Candle . . . There is no better, there is no more open door by which you can enter into the study of natural philosophy . . .

MICHAEL FARADAY, LECTURE 1

Y ou have just stepped down from the omnibus in Piccadilly, and are walking to the west as briskly as possible over the new but already rutted and freezing snow, with the clatter of horse-drawn carriages filling the cold air. Despite the bitter weather, a good number of people are also turning right into Albemarle Street, where, at the far end, stands the Royal Institution. Other members of the audience are descending from their carriages in the street outside, and the entrance is crowded, for today Mr. Faraday is to give another of his Christmas Lectures for young people.

One aim of the Christmas Lectures was to make money for the Institution, which had been established impecuniously by Benjamin Thompson some years earlier. Thompson had fled to London in 1776 when his role as a spy for the British had made his life uncertain in Massachusetts, and soon after his arrival he had sought to establish in England a temple for the propagation of science (as well as a showcase for his own inventions). Another aim of the Institution was to impart to the curious public some of the advances that were then starting to emerge at an accelerated pace from the handful of laboratories around the world, including the Institution's own. The Royal Institution's laboratory was among the few that had been custom-built (in 1799) as a place where scientists could work together cooperatively. One director had been Sir Humphry Davy; now, in the 1850s, the head was his former assistant, Michael Faraday, the largely self-taught son of an impoverished blacksmith.

FARADAY'S INTENTION

When Faraday lived and lectured, the distinction between the branches of science was established but still rudimentary. Although today Faraday is chiefly remembered for his achievements in physics, particularly in the study of electricity and magnetism, he also contributed considerably to chemistry,

Michael Faraday, 1791–1867.

and chose for his discourse on this occasion a subject that illustrated mainly the principles of the latter science. Indeed, although he did not advertise his lectures in those terms, Faraday was in fact introducing his audience to chemical reactions. He selected a candle as the subject of his six lectures, for he knew that the surest way to excite the interest of an audience was to reveal the depths that lie beneath the familiar and the simple.

When Faraday first unleashed his candle's flame, chemistry, too, was still very primitive. It was already divided into its traditional categories of organic chemistry and inorganic chemistry, but physical chemistry, which provides the theoretical framework for the other two, was as yet barely distinct from them.

Organic compounds are all the compounds of carbon other than its oxides (carbon monoxide and carbon dioxide) and carbonates, such as the calcium carbonate of limestone. All other compounds of the elements, including carbon's oxides and the carbonates, are designated inorganic. The division was originally related to the derivation of organic compounds from organisms, for it was thought that living things were essential to their formation. That view persisted through Faraday's early years, and it was not until the German chemist Friedrich Wöhler showed that a characteristic organic compound—urea, which is found in the urine of mammals—could be formed by heating a simple inorganic compound, ammonium cyanate, that there was a general acceptance that a "vital force" was not intrinsic to the formation of organic compounds. That demonstration did not take place until 1828, and the discovery stimulated Wöhler to write excitedly to the influential Swedish chemist J. J. Berzelius: "I must tell you I can prepare urea without requiring a kidney of an animal, either man or dog." Thus died vitalism, and chemistry became whole.

Chemical industry, too, was primitive, but emerging. Today, the industry transforms trillions of kilograms of matter from one form to another each year. However, a major industrial application of chemistry, the derivation of compounds from coal-tar, was being born only as Faraday was speaking, for the first synthetic dye, mauveine, was prepared by W. H. Perkin in 1856. Faraday was at the peak of his powers in the 1820s and 1830s when chemistry was beginning to emerge from its twilight of centuries and well before the sunburst of knowledge that has accompanied chemistry's progress through the present century. Reactions, though, had certainly been known for millennia before the nineteenth century, for chemical reactions first set humanity on its path to civilization: the transformation of ores to bronze and iron are chemical reactions, and the firing of pottery, one of the principal modes by which ancient art has survived and been passed down to us, is another. The development of the reactions that result in ceramics and glass accompanied the development of civilization. The Pantheon in Rome, for example, is one of the most persistent of chemi-

Until Perkin produced the dye mauveine from coal tar, dyes were obtained from natural sources, such as insects, plants, and molluscs. Perkin made the discovery accidentally in the course of an unsuccessful attempt to synthesize quinine.

cal residues, for the concrete of which it is made set to a solid mass by means of a chemical reaction that was contrived by its makers. Faraday, though, was dead before the current tsunami of production and application struck the shores of the twentieth century.

Faraday knew that a chemical reaction is the transformation of matter from one substance—one distinct species of matter—to another. A chemical reaction is essentially the reaction (in the everyday sense of the word) of one substance to the presence of another substance or the reaction of a single substance to changes in temperature or pressure. Faraday's lectures were largely the tracking of the transformations that his candle's wax could undergo once it had been ignited and reactions could ensue. Before it burned, the wax of his candle was one substance, and as it burned it reacted with the oxygen of the air and formed another substance, the gas carbon dioxide, which in turn could go on to form another substance, the solid calcium carbonate. Faraday used his candle and the reactions its kindling entrained as a microcosm of chemistry, to point up in a delicate way the wide variety of transformations that a mere handful of substances could undergo.

The success of Faraday's lectures lay in his selectivity. There is no boundary to the territory of chemistry, and in less masterful hands his audience would soon have been lost in a jungle of variety. Chemical reactions include the ostensibly destructive processes of combustion, when substances such as natural gas, gasoline, fuel oil, and wood react with the oxygen of the air and release energy for our warmth, our motion, or our destruction. They also include, in contrast, the constructive events of nature, such as the process of photosynthesis, in which carbon dioxide that may have been generated by combustion is converted into vegetation by the action of that overlord of terrestrial energy, the sun. Chemical reactions also include the processes by which industry produces transformed matter each year, as metals are won from their ores, petroleum is converted into polymers and pharmaceuticals, and fertilizers are harvested from the air.

Chemical reactions have been taking place upon and below the surface of the earth for millions of years. The upheavals within the cooling globe by which the elements are distributed between the crust, the mantle, and the core are all chemical reactions. The formation of atmospheric oxygen was a result of a sequence of chemical reactions—photosynthesis again—that took the water that had steamed from below the surface of the earth and vented at volcanoes and converted it into that animal-enabling gas. The forces that have molded our landscapes, both its rocky skeleton and its soft covering of soil and vegetation, achieve their changes through the agency of chemical reactions. The germination of a seed is the outcome of many chemical reactions; so is its growth, maturation, reproduction, death, and ultimate decay. The emergence, continuation, and replication of life in all its

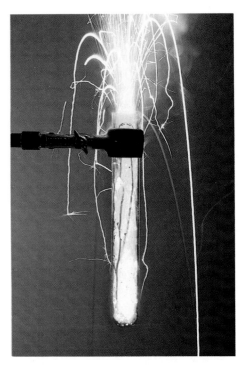

A chemical reaction is the formation of a new substance when a starting material responds to the presence of another substance, or to a change in conditions. In this reaction, the element aluminum is reacting violently with the element bromine, producing aluminum bromide.

CHAPTER ONE

forms are the complex outcome of underlying chemical reactions: wherever we look, even in the act of looking, there is a chemical reaction.

That the panoply of biological and mineral change is an outward display of an inner reaction is an indisputable fact. Some think it sufficient to account for all the rich texture of life, others regard it as merely necessary but not sufficient. Whatever view is taken on that particular point, it is indisputable that the workings of life are chemical and that biological existence is the outer manifestation of intricate reactions. Faraday himself recognized that, for in his lectures he said:

In every one of us there is a living process of combustion going on very similar to that of a candle, and I must try to make that plain to you.

The processes that distribute the elements between the various zones of the planet are chemical; they drag material such as iron to the core and lead to the deposits of metal ores in different regions of the crust. They also generate the gases that contribute to the atmosphere.

THE DEVELOPMENT OF UNDERSTANDING

Faraday was a necessarily primitive scientist (as even the comparable among us a hundred years hence will also seem), for it was only in the eighteenth century that chemistry had begun to develop seriously as a subject, largely under the influence of Antoine Lavoisier in France. Lavoisier and others began to identify the elements—the building blocks of matter, such as hydrogen, carbon, nitrogen, and oxygen—and they started to recognize the laws by which the elements combine. Careful quantitative experiments showed that every compound has a characteristic, fixed composition and that when the same elements combine together to give more than one compound, then the proportions of the elements display a simple pattern of values. The laws of chemical combination were to be the foundation of one of the great evolutionary leaps of the subject, when John Dalton demonstrated in 1807 that the laws of chemical combination pointed to the existence of atoms, the smallest, presumed "uncuttable," particles of elements that could exist. Such is the rationalizing power of this hypothesis that Dalton's atoms have now become the common currency of chemical discourse.

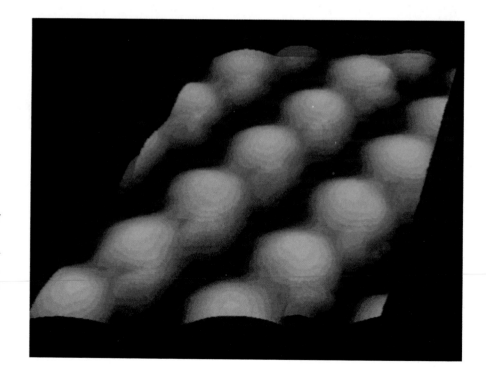

Although Dalton had to infer the existence of atoms from measurements of the masses that combined together, modern techniques can reveal the presence of individual atoms directly. This image of gallium (blue) and arsenic (red) atoms at the surface of a sample of gallium arsenide was obtained by the technique of scanning tunneling microscopy, which is explained later.

Faraday presented his lectures as though for him a reaction were the transformation of substance of one composition into a substance of another composition; he did not present reactions as the modification of an inner structure, as we would today. Partly, no doubt, that was in deference to the youthfulness of his audience. However, it was also partly because some still considered atoms to be mere accounting devices, not actual entities. Although Faraday knew that substances differ in the numbers of atoms of each element that are present (he was one of the best chemical analysts of his day), he would have had little conception of the patterns in which the atoms are linked together to form the molecular structure of a compound. Yet we now know that this structure is often crucial to the distinction of one substance from another. The concept of molecular structure did not emerge until 1858, not long before Faraday's last lectures in 1860. Now we know that two compounds may contain atoms of the same elements in the same proportions, yet because these atoms are linked together differently, the identities of the substances are distinct.

This ignorance of molecular structure meant that Faraday would have been hamstrung if he had sought to explain the processes that he was using his candle to illuminate. A major distinction between his understanding of reactions and our own is that whereas he regarded a reaction as a change in appearance, properties, and composition (although in Faraday's time even composition was difficult to determine and often unknown), we regard reactions as the rearrangement of atoms. Indeed, in some chemical reactions a change of substance is achieved not by change of composition but by changing the pattern in which the same atoms are joined. Faraday knew that combustion converted one substance into another, and he could argue cogently that the carbon and hydrogen of the candlewax combined with the oxygen of the air to give substances that he was able to identify as carbon dioxide and water; but he could not—in any detail—visualize the reactions in terms of the stripping of atoms from one compound and their use in the reconstitution of others. Now we know that a carbon atom (denoted C and represented by a black sphere in the molecular models we show) is carried off by the oxygen (O, represented by red spheres) of the air as carbon dioxide molecules, CO_2, and that hydrogen atoms (H, the small white spheres) are carried off as water molecules, H_2O. Later, when we encounter more complex molecules we shall represent nitrogen atoms (N in formulas) by blue spheres; any other elements we need, we shall introduce as we meet them.

Carbon dioxide and water are among the simplest of all molecules, and little additional knowledge of their properties and behavior can be gained from a knowledge of their structures. However, some molecules are intricate filigrees of atoms, and their structures are crucial to the interpretation

Carbon dioxide, CO_2

Water, H_2O

The chemical reactions that biological molecules perform generally depend crucially on their structure, both the links between atoms and the arrangement of atoms in space. Some diseases and poisonings are the outcome of a minor change in shape that incapacitates the molecule. This computer-generated image is of the enzyme thymidylate synthase. The parts are color-coded to represent the different functional regions of the protein molecule. This enzyme is involved in one of the reactions used to make a component of DNA in living cells.

of their properties. The remarkable union of physics and chemistry that has been so fruitful throughout this century has given us x-ray diffraction, which enables us to discern the detailed structures of the enormous molecules that are so characteristic of biology. Through the technique of x-ray diffraction, we can trace the detailed arrangement in a molecule that consists of thousands of atoms, and sometimes tens of thousands, and through that knowledge of structure, begin to see how that molecule can carry out its role in a living cell.

Some might legitimately argue, even today, that Faraday's ignorance of molecular structure in terms of atoms and of atoms in terms of subatomic particles was simultaneously his strength, for it gave liberty to his considerable originality. Moreover, his demonstrations appealed directly to the minds of his audience because they touched their senses directly. Those present could see the disappearance of the candlewax and the formation of a gas in its place; they could see the formation of clouds of white solid when the gas from the combustion was passed through lime water (a solution of calcium hydroxide—slaked lime—in water). A century and a half later, when we feel impelled to explain reactions in terms of atoms, we are more

When carbon dioxide passes through lime water, a dilute solution of calcium hydroxide in water, the insoluble white solid calcium carbonate forms and comes out of solution. This is one of the experiments that Faraday demonstrated to his audience.

remote from the tangible. Yet this change of focus from the reality of visible form to the unfamiliarity of atoms is the necessary price to pay if we are to penetrate the surface of change, which was all Faraday could observe, and reach the richer dimension of atomic behavior that yields detailed explanations. It is through the atomic, and below that the subatomic, that the tangible becomes explicable.

The knowledge that atoms are the currency of explanation, however, is only half the achievement of the years since Faraday set out to capture the imaginations of his listeners. Just as important for a thorough understanding of the changes that occur in the course of reactions is the shift in the paradigm of explanation that accompanied the replacement of the seventeenth-century classical physics of Isaac Newton by the quantum mechanics that Erwin Schrödinger and Werner Heisenberg introduced in 1926. In quantum mechanics, the distinction between waves and particles fades on the scale of atoms, and probabilities take the place of precise locations and trajectories.

The principles of quantum mechanics are central to a comprehension of the intimacies of reactions. Although it is largely true that the finest chemis-

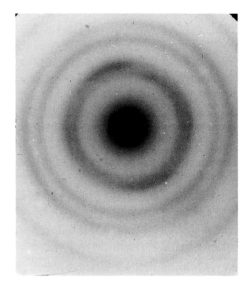

Quantum mechanics is based on the experimental observation that particles can show wavelike properties. One piece of evidence is that electrons, which were thought to be particles, can show the diffraction patterns that are characteristic of waves. This is one of the diffraction patterns G. P. Thomson obtained by passing a beam of electrons through a thin gold foil.

try can be carried out largely in ignorance of quantum concepts, as soon as we seek to delve into the cogs and wheels of chemical change its principles become unavoidable, for matter is intrinsically quantum mechanical. When chemists bring reactant molecules together by the act of mixing and stirring, it is through quantum mechanics that atoms settle into new arrangements, for its principles influence both the location and strength of the new links formed between atoms. When chemists heat a reaction mixture, their purpose is to lever reactants over the energy barriers that prevent immediate reaction, and these barriers are consequences of quantum mechanics. The more we scrutinize the details of reactions, even those of the most mundane kind, the more we shall see that the transformations of matter are material realizations of quantum phenomena.

THE DEVELOPMENT OF TECHNIQUE

The conceptual poverty in which Faraday functioned both in his own research and in his popular expositions was only one fetter on his skill. Another was the crudeness with which he and his contemporaries manipulated matter. They were confined to gross manipulation: to mixing, stirring, and heating—in a word, hoping, and knowing little about what these stimuli were achieving in terms of atoms. This is in marked contrast to the precision that we now know nature can achieve, and which chemists currently strive to emulate. This precision can be seen nowhere more clearly than in the steps by which reactions take place in biological cells. A biochemical reaction can be understood in terms of the *precision* with which events unfold. This is not the mere precision of a modern laboratory, where the success of an experiment may depend on the delivery of a microliter of solution at a particular stage in a reaction, but it is a precision in space and time that is measured in terms of atoms and instants. Thus, in one reaction a single electron—the subatomic particle that we shall see is the key to much of chemistry—may need to be hustled away from a molecule before that molecule has time to react in an inappropriate way that, if allowed, would show itself as disease. A particular atom may need to be removed from one point of a molecule a billionth of a second before another is attached if the reaction is to run a particular course and the cell is not to die. Nature is supremely good at organizing such specificity in time and space, for it can use certain molecules as though they were copper wires to deliver electrons precisely and on time, and it can find ways to bend molecules into particular contortions before sealing the arrangement in a permanent ring. We shall see something of this precision when, at the end of our journey, we discuss the current understanding of photosynthesis, one of the supremely well-orchestrated series of reactions that occur in biological cells.

Chemists currently strive to achieve the fine control exercised by nature. To do so, they still stir, pour, heat, and distil, just as they have done ever since their intellectual ancestors vainly sought a reaction that would produce gold from lead. These crude processes seem to be ways of bending matter to our will and forcing it to undergo specific change. Modern chemists, though, use these techniques to direct reactions more precisely and rationally than alchemists, cooks, and Faraday's contemporaries. They may seek to build a complex molecule, and to do so proceed by stealth and subtlety. Chemists have found ways to take a molecule that is already available and attach an atom to it, close chains of atoms into rings, break open other rings, and gradually elaborate the molecule's structure until it has become the product desired. They have found ways to emulate (and sometimes improve on) nature by mixing, stirring, and heating in such a way that they do not break asunder what they have already joined—even though they cannot manipulate the atoms directly.

Since structure as much as elemental composition is crucial to the identity of a substance, nature and the chemist must contrive to use a reaction to achieve a precise arrangement of atoms, not merely a random pile of atoms of the correct overall composition. Thus the construction must be carried out in sequence, with each atom attached to a precise location on a pre-existing molecular backbone, and then an atom attached to that atom, so that the molecule grows into its final shape atom by atom.

Heat is one potent weapon in a chemist's armory. Heat shakes the molecules, and the violence of the motion breaks them into fragments or causes them to smash together vigorously. For example, molecules of the gas ethylene are obtained by breaking into smaller molecular fragments the heavier hydrocarbons—the compounds of only carbon and hydrogen—that occur in petroleum, and this so-called cracking is achieved through the agency of heat. Heat alone is often too blunt an instrument for today's reactions, however, for it leads to too many unwanted products and thus wastes the precious starting material. So in industry heat is often allied with the utilization of carefully selected catalysts: substances—usually solids, but increasingly liquids, too—that facilitate specific reactions. The deep foundation of industrial chemistry is not so much the marketplace, the plant, or the investment, but the development and selection of catalysts. They are the contemporary philosopher's stone of our subject—devise a catalyst, and you devise an industry. Nature, as might be expected, had arrived at catalysts well before they were contrived by chemists, for living cells contain molecules called "enzymes" that act as biological catalysts, and each enzyme has a specific reaction to facilitate. At the level of chemical discourse, life is an orchestration of catalysis.

Ethylene, C_2H_4

In a "catalytic cracker," large hydrocarbon molecules obtained from petroleum are heated in the presence of a catalyst and are broken into fragments, such as the gases ethylene, C_2H_4, and propylene, C_3H_6, which are used to make the synthetic plastics polyethylene and polypropylene, respectively. This plant "cracks" petroleum to obtain propane.

Ammonia, NH_3

The discovery of an appropriate and efficient catalyst, for example, underlies the production of that queen of synthetic substances, ammonia, NH_3. Ammonia is made almost exclusively by one single procedure, the Haber process, which draws in nitrogen from the atmosphere and combines it with hydrogen obtained from petroleum and natural gas. The process was developed shortly before World War I in a fruitful collaboration between the German chemist Fritz Haber, who developed the catalyst (a form of iron), and his colleague Carl Bosch, the chemical engineer who developed the high-pressure, high-temperature plant that was needed for the implementation of Haber's process (before then, industrial plants had never run under such demanding conditions). An early application of the process was to the production of explosives; they are still one of its destinations, but it is used principally in the manufacture of fertilizers and the plethora of molecules that contain nitrogen, including nylon. Ammonia is nitrogen's gateway to the Vanity Fair of chemistry, for cool, chemically aloof, unreactive, elemental, atmospheric nitrogen, N_2, is brought into the hurly-burly of reactions by its combination with hydrogen: nitrogen is "fixed"—a better term would be *released*—when it forms ammonia, for then it becomes open to ready conversion into a multitude of compounds.

Catalysts are used on a massive scale in industry to bring about chemical reactions that produce polymers, fuels, and all the other derivatives of petroleum. Although the concept of catalysis was known to Faraday, the chemical industry of his time was rudimentary and had not grasped the central importance of opening the gates of chemical reactions by supplying small amounts of apparently unrelated materials, such as the iron that Haber used in the synthesis of ammonia. For Faraday, chemical industry was still the relatively primitive pursuit of change, the manufacture of soap, iron, and glass. The immense contribution of catalysis to our quality of life is largely a twentieth-century phenomenon, for it has enabled manufacturers to work economically, systematically, and on an enormous scale.

Yet, not all reactions are carried out on an enormous scale. Chemists in their laboratories contrive new forms of matter—matter that perhaps exists nowhere else in the universe—by the procedures of *synthesis*, in which larger molecules are built up from small. We shall see much of this centrally important pursuit in later pages, but an example that exhibits the precision of molecule-building a chemist must achieve is that of aspirin, formally acetylsalicylic acid, which has the structure shown in the margin. (The molecule's hexagonal ring of carbon atoms is a recurring motif in organic chemistry: the hexagon itself, specifically C_6H_6, is the hydrocarbon benzene.) The *strategy* for the synthesis of aspirin from the common substance phenol, shown in the illustration on the following page, is to attach a —COOH group to the ring at the carbon atom that lies next to the one to which the —OH group is linked (so forming salicylic acid), and then to modify the —OH group to form the final compound. The question that immediately arises is how the necessary precise degree of targeting is achieved. How can the second group be induced to attach to the phenol molecule at the exact position relative to the —OH group and not elsewhere in the ring, given that the diameter of the ring is less than one billionth of a meter?

We shall see that chemists exercise this degree of control by making use of the properties of the molecules themselves and by an adroit selection of the reagents (the chemicals that are used to modify the structures of the

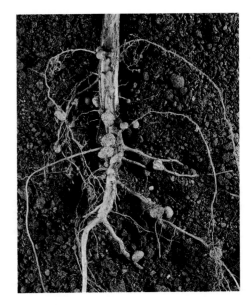

Nitrogen is also "fixed" naturally by the bacteria that inhabit these nodules on the root of a pea plant. Chemists are currently seeking catalysts that emulate the action of the bacteria so that the costly high-temperature, high-pressure Haber synthesis can be avoided. They may need to resort to genetic engineering rather than classical chemical techniques to achieve their goal.

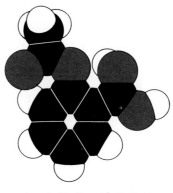

Acetylsalicylic acid, $C_9H_8O_4$

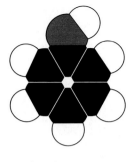

Phenol, C_6H_5OH

starting materials) and the conditions under which the reaction takes place. In the first stage in the synthesis of aspirin, for example, the desired targeting is achieved by passing carbon dioxide through the sodium salt of phenol, and—for reasons that we will explore later—the carbon of the CO_2 molecule seeks out the correct linking position automatically. Chemists do not now generally merely mix a potpourri of reagents and presume that in the debris of the resulting reaction there will lie a few nuggets of the species they require: they contrive the sequential attachment of groups of atoms and organize the reaction so that a precise product is obtained in each stage. Then they may distil or filter to separate a desired intermediate product from a range of others that might also have been produced; restarting with this intermediate product, they edge forward in another step. So it continues until the final product is achieved.

Some reactions are extremely easy to contrive and need little in the way of skill and understanding to execute. Thus, if like Faraday we exhale through lime water, a reaction occurs immediately, and a white solid falls out of solution. If we kindle a flame, the reaction of combustion occurs at once.

Although some reactions occur readily, there are some that we have to strive to prevent. A gentler reaction than combustion, but one that gnaws unceasingly at the very fabric of society, is the pernicious reaction of corrosion. In corrosion, it is not wax but iron that burns, and we see no flame from the slow "combustion" of a steel bridge. Nevertheless, what is actually going on beneath the drop of water on the metal surface is akin to combustion. To recognize that corrosion is a cousin of combustion (and the deep relationship is a point we will establish later) gives rise to the hope that perhaps, since we can extinguish combustion, we can extinguish its cousin corrosion, too. This is the kind of power over matter that comes into our hands with chemical knowledge, for by observing the similarities of ostensibly disparate reactions—and, more deeply, recognizing the equivalence of the reactions that are taking place in terms of atoms and their components— we may be able to stem the incoming tide of corrosion—and, of course, attain more positive goals.

Some reactions require an elaborate series of steps and need to be planned with military precision. The computer is very helpful for this selection of routes between reactants and products. Prior to the 1960s, for instance, the synthesis of naturally occurring compounds—the Everests of achievement in synthetic chemistry, for the compounds are so elaborate— was approached largely unsystematically. However, at about that time the American chemist Elias Corey (who, as these words were being written, received the 1990 Nobel Prize in chemistry for his work) developed a general strategic approach to synthesis, called "retrosynthetic analysis." The strategy of this procedure is to consider the target structure and then to

The corrosion of steel objects by rust accounts for a significant proportion of the gross national product of most countries.

CHAPTER ONE

generate (with a computer) a sequence of simpler ancestor molecules until one is reached that is commercially available. Then the synthesis is achieved by beginning with that simple compound and causing it to undergo the sequence of transformation that led to the target.

At the end of the twentieth century, we see the first indications that someday we may be able to achieve reactions by manipulating atoms directly. A recent advance in technology has given us the technique of scanning tunneling microscopy, in which a needle with a point pulled out to a single atom is moved in successive tracks, like a plow across a field, just above the surface of a solid. In the process, the needle can make out individual atoms—and even the shapes of molecules that lie on the surface—almost literally by touch (in practice, by monitoring the variation in current that flows through the tip as it moves across the surface). One outcome of this technique is that there need no longer be any hesitation in accepting the reality of our discourse: these modern techniques portray individual atoms and molecules with such clarity that doubting philosophers can sleep. But with this technique we can do more. For the atom-sized needle point need not play only the passive role of surveyor of the surface; it can act as shepherd, too, and herd the atoms to lie together. So far, the technique has achieved only the ultimate in modesty in advertising, but the next stage—bringing about a customized individual reaction—cannot be far away.

Back in that winter of the 1850s, as we watch Faraday bringing his lecture to a close with the knowledge available to us then, we are able to think of him as a latter-day Medusa who turned the living into stone: in his hands the original organic candlewax has become a gas that is free to dissolve in the oceans, and thence become the mineral of limestone hills. We also know that there is a richer future open to the carbon dioxide (although the details are as yet a complete mystery), for it may enter the pores of a green leaf, and thence resume its organic role; indeed, one day, it may even become a candle for a later Faraday's kindling. As we come out into the street, our initial sense is one of relief that we are out in the fresh air and away from the increasingly pungent odors within. But the shock of the sudden cold fleetingly sharpens our vision, and as we turn into Piccadilly and the wind catches us, for an instant we wonder what chemists will know about a candle and its carbon one or two hundred years hence. Will they know even more than our famous, learned, entertaining Mr. Faraday?

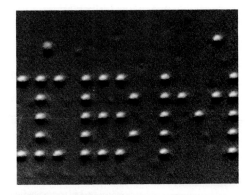

The ability of the scanning tunneling microscope to move individual atoms into predetermined locations is illustrated by this image of xenon atoms that have been arranged on the surface of nickel.

THE ARENA
OF REACTION | 2

*The color of the head of a comet is light
emitted by electronically excited C_2 and CN
molecules; the color of the tail is light
emitted by CO^+ ions. In all cases, the initial
excitation is caused by the impact of
radiation from the sun.*

There is not a law under which any part of this universe
is governed which does not come into play and is
touched upon in these phenomena.

MICHAEL FARADAY, LECTURE 1

Michael Faraday conjured a world of knowledge from a candle. I shall begin where he and the candle left off and attempt to conjure knowledge by pursuing the possible fates of a single atom in its flame. I shall suppose that in the turmoil of combustion there is at least a moment when, somewhere in the flame, there is a carbon atom that has been liberated from the burning wax by the stress of reaction.

What might be the chemical history of that single atom of carbon? Its future might be merely to link to other carbon atoms and form a particle of soot, which would imply at least a brief hiatus in its progress through the world, but its future might also be to form carbon dioxide and to join the gases of the air. There its mobility gives it wings to travel into different forms—perhaps to be absorbed into the oceans, perhaps to lie literally landlocked as limestone until the liberty brought by erosion sends it on its way again, and perhaps to be harvested from the skies by photosynthesis to make that awesome transition from inorganic to organic. One particular atom might in due course become part of a human brain and contribute to the awareness of its own past and future potential. How can a single atom achieve so much? Wherein lies its impressive potential? How, indeed, can that potential be realized?

We shall pursue the atom through the world, seeing how it enters into partnerships and unconscious conspiracies, how it is assaulted by other atoms, forms bonds with them, and disrupts and drives out others. In brief, we shall see a little of the atom's contribution to the ceaseless dance of the elements as they participate in the unending parade of reactions that shape and then reshape our world. In this chapter we concentrate on the *potential* of the atom, the features that make it such a virtuoso of change; in later chapters we shall see how these changes are realized.

The atom at the focus of our interest—carbon—is unique in the sense that it has stumbled into life: the reactions it can undergo, the liaisons it can enter into, are mundane, and they are shared, to a greater or lesser degree, by all the other elements. Carbon's kingliness as an element stems from its mediocrity: it does most things, and it does nothing to extremes, yet by virtue of that moderation it dominates nature. The pursuit of the carbon atom will show us many of the reactions that other elements can undergo, but nowhere in so well poised, so restrained a way.

To understand the reactions of an atom, we must understand the behavior of electrons, the minute, negatively charged particles that surround every atom's nucleus. A carbon atom and the molecules that contain carbon respond to changes in their environment through the malleability of their electron distributions, and particularly through the way in which electrons can be moved from place to place or simply moved aside to expose the nucleus within. The electrons of an atom are the source of its chemical personality, since even a tiny depletion or augmentation of electron density can have a profound influence on the reactions the atom can undergo. So, to see how a carbon atom becomes a particle of soot or a part of a brain, we need to understand how its electrons are distributed and how they may be moved. Such knowledge will give us insight and power. The knowledge may be that when a carbon atom acquires a neighbor—perhaps a hydrogen atom or an oxygen atom—that neighbor may suck away some of the carbon atom's electrons or endeavor to pile more electrons on it. The power is that an enzyme or a chemist may be able to select the neighbor to achieve a particular tiny modification, a particular ripple in electron density, and hence direct the course of an atom's reactions. It is through such tiny, subtle modifications of the distributions of electrons that nature thrives and industry bends matter to its will. The complex reactions of photosynthesis are consequences of tiny ripples of electron density; an industrial plant, big as it is, is designed to achieve particular tiny ripples and to allow their consequences to flow.

CARBON'S ELECTRONS

Faraday could perceive the formation of soot, the luminance of the flame, and the formation of carbon dioxide, and he could discuss them with perspicacity, but a detailed grasp of their mechanism was beyond the reach of his science, the classical physics of Newton. When he made his demonstrations to a presumably enthralled audience, the concepts and language needed to explain them lay almost a hundred years in the future. It is the advance in science that took place in the twentieth century, and particularly the introduction of quantum mechanics, that has made possible a detailed understanding of the chemical personality of an atom. Indeed, all chemical

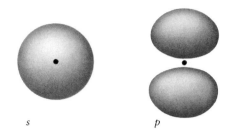

s   p

Two of the boundary surfaces used to represent atomic orbitals: an s orbital (left) and a p orbital (right). The surface encloses the region within which an electron is likely to be found.

reactions, not only those of carbon, are material manifestations of quantum phenomena.

We shall leap across that century of emerging understanding and put our eye to the interior of the carbon atom from a modern point of view, the view that grew haltingly from Ernest Rutherford's vision of an atom as consisting of electrons in orbits about a central nucleus, became quantitatively testable when Neils Bohr proposed that only certain orbits were admissible, and then flourished in the mid-1920s when Erwin Schrödinger formulated his wave mechanics that treated an electron as a wave. The central quantum mechanical idea on which the modern description of the structure of the atom is based is that its electrons occupy "orbitals." Strictly speaking, an orbital is a "wavefunction," a mathematical function that tells us the probability that an electron can be found at each point of space. However, we do not need such coldly quantitative detail, for it is possible to represent an orbital pictorially. To do so, we draw a surface—a "boundary surface"—that surrounds the regions in which the electron is most likely to be found. The orbital can then be thought of as the region of space within this surface. An orbital may belong to a single atom (an "atomic orbital") or—as we shall see in more detail when the carbon atom becomes a molecule—it may spread over several atoms (a "molecular orbital").

Of the several varieties of orbitals in atoms, two play a major role in the chemical reactions of carbon and the other elements that we shall consider (principally hydrogen, oxygen, nitrogen, and the halogens fluorine, chlorine, and bromine). An "s orbital" has a spherical boundary surface, which tells us that an electron with this distribution (an "s electron") will be found anywhere in the sphere bounded by this surface. The dumbbell shape of a "p orbital" shows that a p electron may be found on either side of the nucleus, but not on the plane that divides the lobes of the dumbbell.

Orbitals have a very special, quantum mechanical feature that will prove to be crucial to the behavior of an atom: orbitals are waves in the

A wave in water has alternating regions of positive and negative displacement, and so too does an orbital have regions of positive and negative sign. From now on we represent positive regions by a red region and negative regions by a green region, as shown in the projection below the wave.

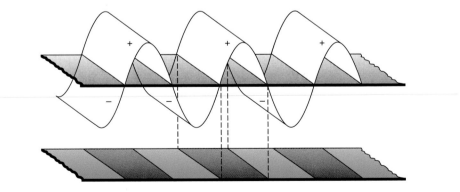

sense that they have regions of positive and negative sign. An analogy can be drawn with a wave in water, which has a "positive" region where the height of the water is displaced upward, and a "negative" region where the water is displaced downward and forms a trough. The positive and negative regions of an orbital have no *direct* physical significance—unlike a wave in water, a positive region of an orbital does *not* mean that an electron is more likely to be found there. As we shall see, however, the regions do have a profound indirect significance and we need to be aware of them. For the time being, we shall treat the sign of a particular region of an orbital as an attribute like "color," and use red for the region where the orbital is positive and green for the region where it is negative. The recognition by quantum mechanics that an electron distribution has "color" is one of the reasons for the success of that theory in explaining chemical reactions: classical physics was "color-blind," and unaware that the distribution of an electron has a "color" as well as a shape.

We should be puzzled by another feature of Faraday's candle. Why, we should worry, does the candle exist as identifiable matter? Why is it that candlestick, candlewax, wick, flame, and all the other atoms of the air do not merge into a single indistinguishable mass, a single cosmic atom? Why, indeed, is the carbon atom in our flame an entity distinct from its surroundings? An answer to such deep puzzles of existence must, apart from their philosophical interest, have a bearing on chemical reactions, for reactions achieve the transformation of one distinguishable variety of matter into another, and we could not speak of atoms if an atom were indistinguishable from its surroundings.

Our current understanding of the reason for the existence of distinguishable entities—why all matter does not simply merge into a cosmic blob—is based on the universal principle discovered in 1924 by Wolfgang Pauli when he was struggling to explain certain features of atomic spectra. His "exclusion principle" states that *no more than two electrons may occupy the same orbital.* That is, no more than two electrons in any given atom can occupy any one of the atom's *s* orbitals, no more than two electrons can occupy any one of the atom's *p* orbitals, and so on. One consequence of the exclusion principle is that electrons belonging to neighboring atoms cannot occupy the same region of space, so the atoms cannot merge. Thus, Pauli and his principle render the candlestick distinguishable from the candle. Moreover, a carbon atom acquires its bulk from the principle, for its electrons cannot all occupy the orbital of lowest energy, but must occupy the concentric shells of orbitals that surround the nucleus. Shortly we shall see precisely which orbitals are occupied, and how the occupation of different orbitals in atoms of different elements leads to their different chemical properties.

A p orbital with the signs of the distribution represented by colors: as in the preceding illustration, red denotes positive regions and green represents negative regions.

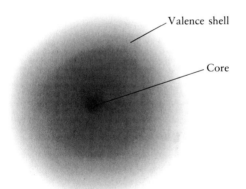

Valence shell

Core

Whereas Faraday might have thought of a carbon atom as a solid sphere, our current understanding is that it consists of a nucleus and a core of two electrons surrounded by a more diffuse shell of four electrons. The density of electrons is high in the core (darker gray), and decreases farther from the center (lighter gray).

The Pauli exclusion principle is in fact an even stronger constraint on electrons than we have suggested, for it goes on to make a statement about two electrons that occupy the same orbital. This feature depends on the fact that an electron has the property known as "spin." For our purposes we can safely imagine spin to be a rotational motion of the electron on its axis. (In fact it is not a motion of the kind familiar from classical physics but a purely quantum mechanical phenomenon that has certain bizarre—nonclassical—properties.) An electron "spins" at a fixed, invariant rate, and all electrons spin at exactly the same rate. The rate of its spin is just as much an intrinsic property of an electron as its mass and charge, which are also identical for all electrons. However, according to quantum mechanics, an electron may spin in one of two directions: it can have clockwise spin (denoted ↑) or counterclockwise spin (denoted ↓).

As well as limiting the occupancy of any orbital to two electrons, the Pauli exclusion principle goes on to assert that if two electrons do occupy any one atomic orbital, then they must spin in opposite directions: one must be ↑ spin and the other ↓, resulting in zero net spin within each orbital. Classical mechanics knew nothing of spin; we shall see, however, that through the Pauli exclusion principle spin is intrinsic to the chemical properties of matter. There was no hope for Faraday: his classical world was wholly unequipped to understand the chemical history of a candle (but, how well he did without understanding what he did!).

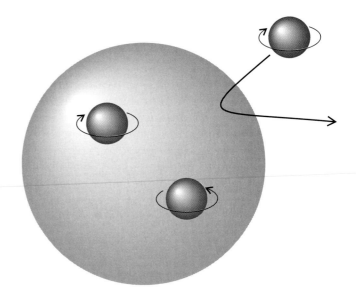

According to the Pauli exclusion principle, an orbital can contain at most two electrons, and those two electrons must have opposite spins. The exclusion principle is responsible for the bulk of matter, for it prevents large numbers of electrons occupying the same region of space.

CHAPTER TWO

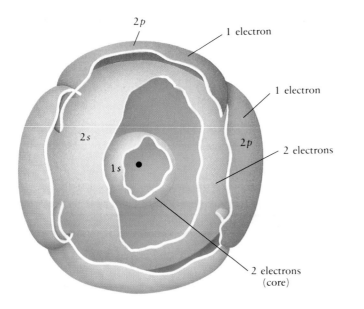

2p

1 electron

1 electron

2s

2p

2 electrons

1s

2 electrons
(core)

A much more detailed picture of the structure of the carbon atom than was available in Faraday's time. Two electrons occupy the 1s orbital and form the core of the atom. Two more occupy the 2s orbital with opposite spins, and complete that orbital. The remaining two electrons occupy two separate 2p orbitals.

Now when we look through the flame at the carbon atom, we can see it with detailed understanding. A carbon atom has six electrons. Because no more than two electrons can occupy any one orbital, the six electrons are arranged in a series of concentric shells, a little like the layers of an onion. The innermost shell of the atom consists of a single *s* orbital—since it is an orbital (indeed, the only orbital) of the first shell it is called a 1*s* orbital—with its two electrons. Around that innermost shell is a second shell, containing two electrons in a 2*s* orbital. The second shell also has three *p* orbitals (the 2*p* orbitals), and the remaining two electrons occupy two of them (with one electron in each). The six negative charges of the electrons balance the six positive charges of the carbon nucleus, so overall the atom is electrically neutral.

It will turn out to be a central unifying principle of chemistry that the second shell of orbitals can accommodate up to eight electrons (two electrons in each of its four orbitals, one 2*s* and three 2*p*). A typical atom—like carbon—consists of one or more complete inner shells of electrons and an incomplete outer shell of several electrons that surround that inner "core" of electrons. The electrons in the outermost shell are called the "valence electrons," for they are largely responsible for the "valence" of the atom, its ability to form chemical bonds that link the atom with other atoms when reactions create new forms of matter. "Vale!" ("Be strong!") was what the

Romans said on parting, and the crucial aspect of a chemical bond is its strength.

The valence electrons are the electrons that determine whether a bond is strong, whether it may be formed, and whether it may be broken. In virtually all chemical transformations, the valence shell of an atom is the center of the action in a reaction. For the carbon atom in the flame, it will be the four valence electrons that determine its destiny.

THE FORMATION OF BONDS

Chemical reactions are processes in which old bonds between atoms are severed and replaced with new bonds to different atoms. Faraday's candlewax is composed of molecules that are largely composed of strings of linked carbon atoms, each one about 30 atoms long, with either two or three hydrogen atoms attached to each carbon atom. When the wax travels up the wick and burns, the bonds between the carbon atoms and between the carbon and hydrogen atoms are broken apart. Such is the stormy turmoil in the conflagration—particularly on an atomic scale, where great forces are unleashed to produce even a gently flickering flame—that the molecules are ripped apart into fragments and even into individual atoms. Much later we shall see some of the details of this ripping apart of molecules, but for now we concentrate on two possible fragments: a carbon atom and a hydrogen atom. We shall suppose that these atoms are destined to combine, and explore what it means for CH and C_2 fragments to form in a flame.

Atoms form bonds with each other either because one atom transfers an electron to another or because the two atoms share electrons with each other. For the arch-compromiser carbon, the major mode of bond formation is the latter. However, it will be helpful to see why it adopts this middle way by considering the former type of bonding first, which is shown by chemically more alert elements, such as sodium and chlorine.

An atom that has only one or two valence electrons can give them up because they are only weakly held by the nucleus buried inside the electron core. When electrons have been lost, the positive nuclear charge is no longer canceled by the negative charge of the electrons, and so an atom that loses electrons becomes a positively charged "ion," a "cation" (Faraday's term). Sodium is chemically alert in this sense and tends to form cations, for it has a single valence electron outside a core of 10 electrons; calcium is similarly alert, and has two electrons outside a core of 18 electrons.

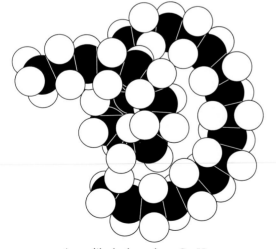

A waxlike hydrocarbon, $C_{30}H_{62}$

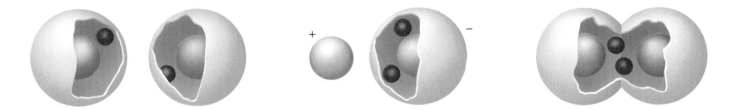

Two atoms (left) can form a bond in two alternative ways. One way (center) is for an atom to transfer an electron into an incomplete orbital on the other atom. The atom that loses the electron becomes a cation (an ion with a positive charge) and the atom that gains one becomes an anion (an ion with a negative charge). The two ions are attracted together by their opposite electric charges. Alternatively (right) the two atoms may share the two electrons, and link together by forming a covalent bond.

The fact that an element can readily give up its electrons is a sign that it is metallic, for metals consist of an array of cations pervaded by a mobile sea of electrons lost from the valence shells of the parent atoms. When we see ourselves in a mirror, we are seeing the oscillations of the mobile electrons in the metallic backing. When we hammer a metal into shape, we are pushing cations past each other through the compliant sea of electrons. When a metal reacts—when sodium sizzles in water or iron quietly corrodes—it is surrendering its electrons. The loss of an electron can transform the character of an atom beyond recognition, for a compound such as sodium chloride (table salt), which contains sodium ions, is utterly unlike its parent sodium metal. It is this transformation of a personality on the loss of even a single electron from an atom that underlies the richness of qualities that can be spun from reactions, for the transfer of an electron may transform a substance entirely. The havoc of rusting is merely the loss of three electrons.

The electrons that atoms such as sodium can release may be transferred to another atom if that atom has an incomplete valence shell to receive the incoming electrons. The atom that gains electrons becomes a negatively charged ion, an "anion." Accepting an electron also transforms a substance's chemical personality: when chlorine reacts with sodium to form sodium chloride, its acceptance of an electron results in its transformation from a poisonous, pale green gas to a colorless solid that is an essential component of our diet.

After a metal atom transfers one or two electrons to another atom, the attraction between the opposite charges of the newly formed cation and anion attract each other and hold the two ions together in an "ionic bond." In fact, large numbers of ions can clump together under the influence of their mutual attraction. The result of the reaction between sodium and chlo-

The silvery metal sodium reacts with the pale yellow-green gas chlorine to produce the white crystalline solid sodium chloride. Both reactants are dangerous chemicals, but sodium chloride is ordinary table salt.

Sodium has one electron outside a core, and readily gives it up to other species. That is one of the reasons why it reacts so vigorously with water to produce hydrogen gas.

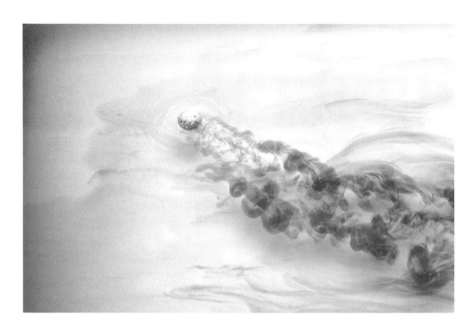

rine is a crystal of sodium chloride, which consists of an aggregate of huge numbers of sodium cations and chlorine anions.

Energy is the criterion that determines whether a pair of elements can participate in ionic bonding. Energy is always needed to detach an electron and form a cation; it is also sometimes needed to attach an electron to an atom to form an anion. The overall energy required will be small if the electron is being detached from a chemically alert metal and is being inserted into an incomplete valence shell of another atom. Bonding will occur if the formation of a bond ultimately releases more energy than is invested. If the overall energy requirement is small, far more energy than is needed to form the ions may be recovered if the ions attract each other strongly, which they do if they are so small that their charges are not far apart.

Carbon and other nonmetallic elements release electrons to other atoms only with reluctance, for their electrons are gripped tightly by the highly charged nucleus at the center of the atom. Although they cannot form cations readily, they can *share* electrons with their partners to form "covalent bonds." A covalent bond consists of two electrons located largely between the two nuclei; the electrons act as a kind of electrostatic glue that holds the atoms together. Covalent bonds are the cement that binds atoms into distinguishable molecules. Thus, the hydrogen molecule H_2 consists of two protons—the nuclei of hydrogen atoms—held together by a pair of electrons lying predominantly between them. We can write the hydrogen molecule as H—H, the bar signifying the shared pair of electrons. The CH fragment that forms transiently in the flame is a carbon atom and a hydrogen atom held together by the pair of electrons they share, and the bond is written C—H. Almost all the species that we shall meet—and all the fabrics, food, gases, and pharmaceuticals of everyday life—are covalently bonded molecules.

In the covalently bonded C_2 fragment, the electrons are equally shared between the two atoms, but in the CH fragment they lie slightly closer to the carbon atom. The evenness with which the pair is shared is influenced by the "electronegativity" of an element. This concept was first introduced by the American chemist Linus Pauling, and was defined by him as the power of an atom of the element to attract electrons to itself when it is part of a compound. High electronegativity is a consequence of high nuclear charge coupled with small atomic size, for the combination of both properties conspires to attract electrons. Fluorine is the most electronegative atom, and in general the elements of high electronegativity are those near it in the periodic table (specifically, nitrogen, oxygen, and chlorine). Shifts of electron density that arise from differences in electronegativity are usually very slight (particularly in the case of C—H bonds, for the electronegativities of the two elements are not very different). Nevertheless, as we shall see, great consequences flow from tiny accumulations or deficiences of electrons.

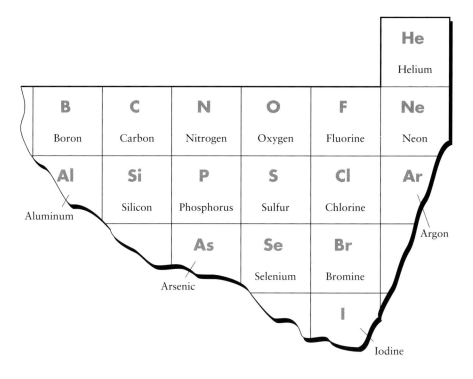

A fragment of the periodic table widely used to display regularities and similarities between elements: elements near each other (particularly those in the same column) have similar properties. Most of the elements we consider are in the region of the periodic table shown here.

The electronegativity of carbon is another measure of the element's potent mediocrity, for its value is broadly midway between that of the most aggressive seeker after electrons (fluorine) and the most virile giver of electrons (the vigorously reactive metal cesium). A carbon atom makes few demands electronically on its neighbors in a molecule: it does not pull electrons off particularly aggressively, nor does it load them on. Carbon is an atom for the quiet life.

The number of covalent bonds that an atom can form is limited by the number of orbitals and electrons in its valence shell. A hydrogen atom (which has only the $1s$ orbital for its valence shell) can accommodate only two electrons before the shell is full, and hence can form at most one covalent bond. This is why a hydrocarbon molecule like that in candlewax is studded with hydrogen atoms, for once a hydrogen atom has attached to a carbon atom in the chain, its capacity for bond formation is exhausted. Carbon (together with nitrogen, oxygen, and fluorine) has s and p orbitals in its valence shell: it can accommodate eight electrons and form up to four covalent bonds with other elements. As a result, carbon can form chains and networks as its atoms link together in complex patterns. A diamond is the apotheosis of this ability to form four bonds, for a single diamond is like a huge molecule in which every carbon atom is joined to four neighbors lying at the corners of a tetrahedron, and the crystal has a structure like a rigid three-dimensional scaffolding. In more typical molecules, the carbon atoms form smaller chains and rings, but every one of them forms four bonds. One possible future for our particular carbon atom is to combine with oxygen. An oxygen atom has six electrons in its valence shell, and needs two more electrons to complete its shell. It can acquire the electrons by sharing electrons supplied by two other atoms, as in the formation of HOH, water (one product of the flame), or by sharing two electrons with one other atom. Note again how the formation of a new bond—the reorganization of the pattern in which electrons are paired—can transform the character of a substance, as when hydrogen atoms are plucked from the hydrocarbon by oxygen and water is formed. When we come to account for massive change, we may find it in the relocation of an electron.

A general rule of thumb in chemistry, which is known as the "octet rule," states that atoms undergo bond formation until they have captured eight electrons for their valence shell (and two in the case of hydrogen). The rule works quite well for the elements we shall meet (which include carbon, nitrogen, oxygen, and fluorine), which have four orbitals in their valence shells, and hence can accommodate up to eight electrons. However, for other elements the rule may be transgressed in one direction or the other, either by an atom swelling its valence shell to accommodate more bonds or by a molecule failing to complete its valence shell. Thus, some atoms (among them silicon, phosphorus, and sulfur) are big enough for up to six atoms to pack around them and to find a means of attaching quite stably (the compound sulfur hexafluoride, SF_6, which is stable enough to be used as a gaseous insulator in electric switchgear, is an example). Other atoms are so small that it is only marginally advantageous for them to acquire eight electrons, and they may be content with six. Thus, the small boron atom has only six electrons in its valence shell in the compound boron trifluoride, BF_3.

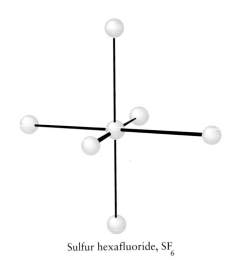

Sulfur hexafluoride, SF_6

Although these byways of bonding may appear too recondite to be of practical importance, that is in fact not the case. Boron trifluoride, for example, is made in considerable amounts by chemical industry. As we shall see, a peculiarity of structure—such as having an incomplete octet—may be turned to advantage by a chemist who seeks to bring about an unusual modification of structure or who wishes to facilitate an otherwise sluggish reaction. Boron trifluoride's chemical charms will start to become apparent in Chapter 3. More to the point at this stage is that carbon has as neighbors in the periodic table the elements boron and silicon: carbon is on the edge of being an atom that can be electron-bloated (silicon) or that can be electron-meager (boron).

When our carbon atom combines with another it may do so by sharing more than one pair of electrons, and forming "multiple bonds." A pair of atoms may share one, two, or three pairs of valence electrons, forming a single, double, or triple bond between them. The C—H bond in the CH molecule and both O—H bonds in an H_2O molecule are single bonds; however, there is a double bond between the carbon and oxygen atoms in carbon dioxide, which is denoted O=C=O when we want to show its bonding pattern. The C_2 molecule is C=C. It is always the case that atoms are bound together more strongly by a multiple bond than by a single bond. The multiple bonds of the nitrogen molecule, N≡N, are the principal reason for that element's considerable inertness, and why it can act as an unreactive diluent of the dangerous oxygen of the atmosphere.

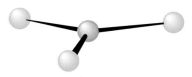

Boron trifluoride, BF_3

There is another peculiarity of carbon in this connection. That apparently innocent atom hanging in the flame has a trick up its sleeve: a C=C double bond is not as strong as two C—C single bonds. This fact is essential to the comprehension of many of the reactions of organic compounds, for it may turn out to be worthwhile (in the sense that a lower energy can be achieved) if an organic molecule bursts one of its multiple bonds and replaces it by several single bonds. Indeed, carbon–carbon multiple bonds are very tender regions of organic molecules, and if our atom ends up attached to a molecule by a multiple bond, it may be open to all kinds of chemical abuse.

That single and multiple bonds have a reality beyond a chemist's musings is shown by our everyday act of burning a fuel, and in particular Faraday's demonstrations with a candle. The heat of a flame is a manifestation of the reorganization of bonds: the turbulent maelstrom of a conflagration is the outward manifestation of the merest shifts of electrons. Something of the changes that take place can be appreciated by considering the combustion of methane, CH_4, the principal component of natural gas and a close analog of candlewax. In this reaction, a molecule of methane combines with two oxygen molecules, O_2, present in the surrounding air. The net change in energy that accompanies the reaction can be set out on the following balance sheet of profit and loss, where profit represents the energy released when a particular bond is formed, and loss represents the energy that must be supplied to achieve the cleavage of a bond (the numbers come from separate measurements of the strengths of bonds):

	LOSS, KILOJOULES		PROFIT, KILOJOULES
Cleavage of 4 C—H bonds	1648	Formation of two C=O double bonds	1486
Cleavage of two O=O bonds	992	Formation of 4 H—O bonds	1852
Total:	2640		3338
		Net profit:	698

The overall transaction is energetically profitable, for although a considerable amount of energy must be invested to separate atoms from their neighbors, the return on the investment is substantial when the liberated atoms settle into new partnerships. A major source of the energy of combustion is the great strength of the C=O double bonds, which stems from the small size of the two atoms and the ability of the electrons that bind them to interact strongly with the two nuclei. Thus, when Faraday ignited his candle, the luminosity of the flame and the warmth it generated was the liberation of energy as heat as a result of this reorganization of the partnerships of atoms, and particularly as a result of the replacement of carbon–hydrogen and carbon–carbon bonds by their stronger carbon–oxygen and oxygen–hydrogen analogs.

LONE PAIRS

Not all valence electrons need participate in bond formation. A fluorine atom has seven valence electrons, but when two fluorine atoms combine to create a fluorine molecule, F_2, only two of the total of 14 valence electrons are used to form the connecting bond. The remaining 12 valence electrons

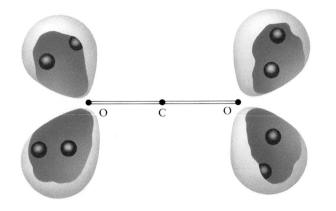

are present as six "lone pairs," three on each atom. If we want to emphasize the presence of lone pairs, we write the structure of the molecule as $:\!\ddot{F}\!-\!\ddot{F}\!:$, which shows each nonbonding electron as a dot.

Lone pairs are far from being innocent bystanders in a molecule. By noticing the presence of lone pairs on the fluorine atoms, for example, we can immediately begin to see one of the reasons for the extraordinary reactivity of fluorine gas, a reactivity that made it more of a laboratory curiosity until halfway through this century when its importance for the refining and treatment of uranium overcame chemists' reluctance to handle it. Their increasing familiarity with the material led to the discovery of several important series of compounds, among them the generally environmentally evil chlorofluorocarbons (which are used in aerosol sprays, refrigerators, and air conditioners) and the helpful polytetrafluoroethylene plastics (such as are used on nonstick kitchenware).

It is easy to understand why an F_2 molecule springs apart into highly reactive fluorine atoms so readily: fluorine atoms are small (they have only nine electrons each, and the nuclear charge grips them tightly), and so the fluorine atoms must be close together if they are to bond at all. However, the closeness of the atoms brings the lone pairs on the neighboring atoms so close to each other that they repel each other strongly. Hence, the molecule is like a compressed spring, and the atoms readily fly apart. The chlorine molecule Cl_2 has a very similar structure, $:\!\ddot{Cl}\!-\!\ddot{Cl}\!:$, with three lone pairs on each atom, but is much less reactive. Because a chlorine atom is much fatter than a fluorine atom, the Cl—Cl bond is significantly longer and hence the lone pairs repel each other less.

A lone pair can take a direct part in a chemical reaction. In brief, a lone pair is an electron reservoir that can be induced to participate in bonding: a lone pair on an atom is a spearhead that can seek out a region of positive charge in a molecule and form a bond with the atom that carries that charge. A part of the strategy of reactions in organic chemistry, for example, depends on electrons drawing back from the vicinity of a specific carbon atom and then the lone pair on an atom of another group moving toward the exposed nuclear charge, like a fish to bait.

MOLECULAR ORBITALS

The picture we have given of a covalent bond as an electron pair confined to the region between the atoms it links is very primitive: it was the one proposed by the great American chemist G. N. Lewis in 1916, well before quantum mechanics had been formulated. In fact, electrons are not confined strictly to a single pair of atoms but occupy orbitals that generally spread throughout the molecule. When we say that an electron pair is located between the two nuclei, as in the CH fragment, we really mean that it is dispersed throughout the molecule, but with a high probability that it will be found between the two nuclei.

One important consequence of this "delocalization" of electron pairs is that it allows influences to be transmitted throughout the molecule, and the

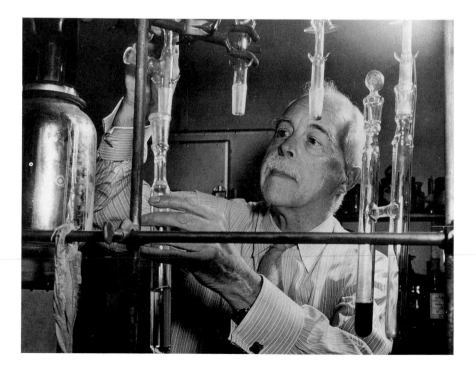

Gilbert Newton Lewis (1875–1946).

replacement of an atom in one location by another atom may have an effect several bond lengths away. Once again, we are seeing that chemistry is subtle, since the activities at the reaction center are likely to be modified by events far removed in a distant part of the molecule.

Incidentally (but crucially), we can now see why the electron pair is the unit of construction of chemical bonds. We have seen that the Pauli exclusion principle limits the number of electrons in an orbital to two. The principle also applies to molecular orbitals: only two electrons can enter a single molecular orbital however widely it is delocalized. Hence, the electrostatic glue consists of two electrons in an orbital that binds the molecule together. Just as in an atomic orbital, those two electrons must be paired in the sense that they must have opposite spin. Thus, when a covalent bond forms, one electron in the bond has an ↑ spin and the other has a ↓ spin. Most compounds have electrons that are fully paired, so if we were interested only in appearances, no sign of the spin is apparent (rather like the blindness of classical mechanics to the "color" of an electron distribution). However, as soon as we need to think about the details of bond formation, we need to take into account that a bond is composed of two canceling spins.

Molecular orbitals can be represented by boundary surfaces, just like atomic orbitals. If we wanted to construct the molecular orbitals for the CH and C_2 molecules—or any of the more elaborate molecules that we meet later—then we would need to solve the Schrödinger equation, the same equation as we solve to find the orbitals of individual atoms but now expressed in terms of all the electrons and nuclei that are present in a molecule. However, because such computations are numerically very demanding, chemists have devised a qualitative procedure called the "LCAO technique" (where the initials stand for *linear combination of atomic orbitals*), which allows them to construct reasonably accurate approximations of the orbitals without going through detailed computation. The arguments they use are central to an understanding of molecular structure, and hence they are central to a comprehension of chemical reactions.

When Faraday gazed so intently at his candle flame, and let it stir his imagination, he was unable to visualize—except as microscopic whizzing and ricocheting balls—the structure of the entities he was describing; nor would he, I think, have been comfortable with the intensity of mathematics that must be invoked in current detailed descriptions of molecular structure. However, I suspect that he might well have relished the qualitative mode of description of molecular structure that chemists have developed as a vehicle for their own insight. He would have been able to gaze into his candle's flame and *imagine* the structural features of the molecules that were forming within it and participating in its luminosity.

Faraday would have little difficulty with the construction of approximate molecular orbitals using the LCAO technique, for this technique

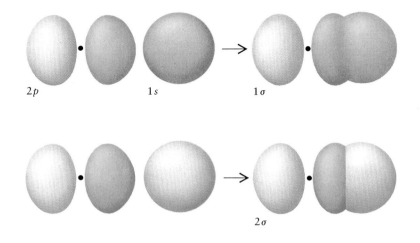

The formation of bonding (1σ) and antibonding (2σ) molecular orbitals when a carbon 2p orbital overlaps constructively (top) or destructively (bottom) with a hydrogen 1s orbital.

renders molecular orbitals open to ready visualization. In the LCAO approximation, molecular orbitals are constructed by superimposing the valence-shell atomic orbitals of the contributing atoms. Suppose, for instance, we needed to construct the orbital that is involved in bond formation in the CH fragment. We would picture one 2p orbital of a carbon atom and its one electron, and the 1s orbital of a neighboring hydrogen atom with its one electron. We could then think of these two wavelike regions as merging—the technical term is "overlapping"—and forming a composite orbital that spreads over both atoms and is occupied by the electron pair.

This description of CH is incomplete, however, because we have not accounted for all the orbitals formed by superimposing the atomic orbitals that contribute to the bond. A general feature of the LCAO procedure is that from N atomic orbitals, N molecular orbitals can be formed. Thus we are missing an orbital, because in CH $N = 2$. To understand how such an orbital arises, we need to examine a *quantum mechanical* point that will prove to be the key to understanding the course of numerous reactions. As we have remarked, quantum mechanical ideas are crucial to a deep understanding of chemical reactions (but not, admittedly, necessarily to their implementation in laboratories and kitchens, where practicing and unwitting chemists, respectively, do very well without all the paraphernalia of quantum theory).

The two overlapping atomic orbitals, like any waves that overlap, may augment or cancel each other depending on whether peak overlaps peak or peak overlaps trough. The two possible superpositions we denote 1σ (the result of constructive interference between regions of the same "color") and 2σ (the result of destructive interference between regions of different "color"). The 1σ orbital, because of the quantum mechanical wavelike interference, is strongly enhanced in the internuclear region, so electrons that occupy it are in a good position to interact favorably with both nuclei and help to draw them together. In contrast, the 2σ orbital is strongly diminished in the internuclear region, because in that region the two overlapping waves cancel each other. This cancellation is complete midway between the nuclei, and is analogous to a line of zero displacement of water where two ripples merge. The electrons that occupy the 2σ orbital are excluded in large measure from the internuclear region, and hence are not in a good position to interact with both nuclei and bind them together. The existence of the node—the surface on which the orbital is zero—in 2σ is a purely quantum mechanical phenomenon since it depends on the cancellation of the wavelike atomic orbitals. The fact that regions of the same "color" interfere constructively but those of opposite "color" interfere destructively and lead to nodes is a pivotal feature of modern explanations in chemistry.

In summary, the overlap of two atomic orbitals results in the formation of the two molecular orbitals of CH, one with an internuclear node and the other without. The two orbitals differ in energy, with the nodeless 1σ orbital lower in energy than 2σ. Both bonding electrons of the CH molecule (one supplied by each atom) can be accommodated in the lower energy 1σ orbital so long as their spins are paired. The simplest explanation of the difference in energy between the two orbitals is that because an electron that occupies the 2σ orbital is excluded from the region between the nuclei, it has a less favorable interaction with the nuclei than an electron in the 1σ orbital, which can be found in the internuclear region. Because internuclear electrons hold molecules together and electrons that are found largely outside the internuclear region tend to pull molecules apart, the energy-lowering 1σ orbital is called a "bonding orbital," and the energy-raising 2σ orbital is called an "antibonding orbital."

A modern day Faraday would now see in his imagination two paired electrons in the sausage-shaped 1σ orbital formed by the constructive overlap of the two orbitals of the neighboring atoms. The two electrons in a 1σ orbital are called a "σ bond." The common feature of σ orbitals (both bonding and antibonding) is that they all have cylindrical symmetry around the internuclear axis (and hence resemble s orbitals when viewed along that axis).

Now consider the C_2 molecule that our carbon atom might form if it collides with another carbon atom, or that may result if a fragment of the hydrocarbon chain is stripped of its hydrogen atoms in the conflagration. Each atom of this species is richer in orbital structure than a hydrogen atom, and so we can expect that the two atoms will be linked together by a more elaborate system of bonds. In fact, the two atoms share three electron pairs in three different orbitals. Two of these orbitals have electron distributions that are strikingly different from those of σ bonds, and it is the purpose of

The eight molecular orbitals that can be formed from the eight atomic orbitals in the valence shells of two carbon atoms. The energy of the orbitals increases up the page, with 1σ the lowest energy (most strongly bonding) and 4σ the highest energy (most strongly antibonding). In the ground state of the molecule, the 1σ, 2σ, and 1π orbitals are occupied by paired electrons.

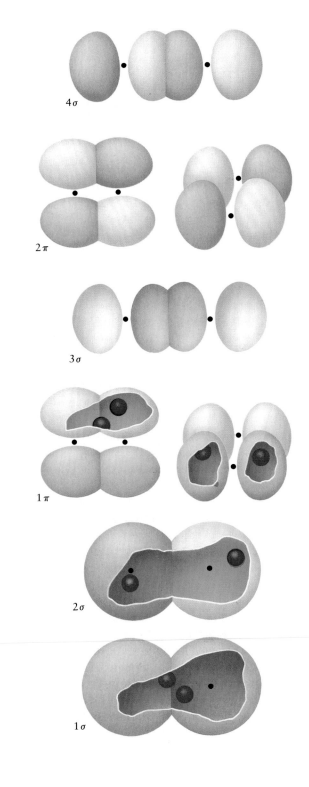

this excursion into C_2 to introduce them, for much will spring from them later.

We know that each carbon atom has a $2s$ and three $2p$ orbitals in its valence shell, so we must explore the patterns that form when these are superimposed. It is simplest to consider the superposition of similar orbitals on each atom ($2s$ with $2s$, and so on). First we consider all the orbitals that can be formed, which are shown in the illustration, and then look at which of them will be occupied by electrons.

The two $2s$ orbitals overlap to form bonding and antibonding σ orbitals (1σ and 2σ). The two $2p$ orbitals that point toward each other also overlap with the cylindrical symmetry characteristic of σ orbitals, and they form a bonding and antibonding combination of their own (3σ and 4σ). However, the two $2p$ orbitals on each atom that are perpendicular to the C—C axis overlap to give shapes that are different from the ones we have seen so far. As each combination resembles a p orbital when viewed along the axis, the four combinations, two bonding and two antibonding, are called "π orbitals." Each π orbital can accommodate two paired electrons. We denote the two lower-energy, bonding combinations as 1π orbitals (the two have the same energy) and the two antibonding orbitals as 2π orbitals (they too have the same energy as each other).

We can now picture the structure of the C_2 fragment in terms of the orbitals we have just described. We have eight electrons to accommodate, and they enter the orbitals with the lowest energy subject to the restriction of the Pauli exclusion principle. Two of them pair, enter, and fill the 1σ orbital, and two more pair, enter, and fill the antibonding 2σ orbital. The next four form two pairs, and enter the two 1π orbitals, and fill them. A modern Faraday's vision of the molecule illustrated at the top of the page would therefore be a series of shells—not greatly unlike a stretched-out atom—with a tight core of electrons close to each nucleus, then a sausage-shaped σ orbital with plenty of electron density between the two nuclei, and another (antibonding) σ orbital with electron density outside the internuclear region. Finally, he would visualize two π-orbital distributions in regions parallel to but outside the internuclear region.

When we see the blue glow of gas flame, we are watching the C_2 molecules that we have just described. When a C_2 molecule is first formed in the tempest of the reaction, its electrons do not have their final distribution and the molecule is in an energetically excited state. However, the molecule quickly collapses into its state of lowest energy, or ground state, and as it does so the readjustment of its electrons gives an impulse to the electromagnetic field. This shock generates a photon of light that carries off the discarded energy. In the case of C_2, the molecule generates a photon of blue light, which is what we see. So, when we watch a blue gas flame, we can imagine the light as emanating from the C_2 molecules as they discard their

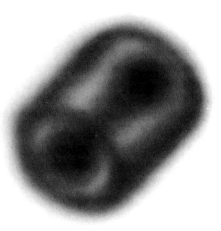

A depiction of the shell structure of the C_2 molecule. The two 1π orbitals give rise to a cylinder of electron density around the axis of the molecule. The electron density on the axis arises from electrons in the 1σ and 2σ orbitals, which jointly make little contribution to the bonding.

energy and settle into their final electronic states. The same radiation is detected from the head of a comet, and is accompanied by violet and red light from similar transitions in a CN fragment. (The light from the tail originates from CO^+ ions.) Comets are not alight, of course, and in their case the light is emitted from molecules that have been excited into higher-energy electronic arrangements by the impact of solar radiation.

Back to earth, from comet to candle, and the description of molecular structure. We can interpolate a remark at this stage: what previously we have called a "double bond"—two shared electron pairs—can now be interpreted in molecular orbital terms as consisting of one σ bond (an electron pair in a bonding σ orbital) and one π bond (an electron pair in a π orbital). We can also see the explanation of why a double bond does not necessarily have the strength of two single bonds: the distributions of electrons in its σ and π components are quite different from each other and a $\sigma+\pi$ combination is not the same as two σ bonds.

The blue color typical of a flame fueled by hydrocarbons (here natural gas, which is largely methane, CH_4) burning in a plentiful supply of air is generated by electronically excited C_2 molecules discarding their energy as electromagnetic radiation.

MOLECULAR ORBITALS FOR LARGER MOLECULES

The same principles of superposition and constructive and destructive interference apply to larger molecules as to smaller ones, the only difference being that atomic orbitals from all the atoms contribute to each molecular orbital, and the resulting orbitals spread throughout the molecule. We can see what is involved by looking slightly into the future of our carbon atom and constructing a molecular orbital description of the carbonate ion, CO_3^{2-}, that it might one day form if carbon dioxide were to dissolve in water. The carbonate ion is a planar species in which a carbon atom lies at the center of an equilateral triangle of oxygen atoms. We use the "Lewis structure" of the ion to show its shared and lone electron pairs, and in this notation the ion is depicted as a blend (the technical term is a "resonance hybrid") of the three structures

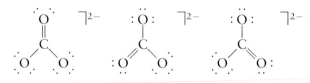

and we see that the oxygen atoms are bound to the central carbon atom by a mixture of single and double bonds. The bracket with 2− on it signifies that the double negative charge of the ion is spread over all four atoms.

To keep the discussion within bounds, we shall concentrate on constructing the π orbitals of the ion, which are built from the overlap of carbon and oxygen $2p$ orbitals that lie perpendicular to the molecular plane

Carbonate ion, CO_3^{2-}

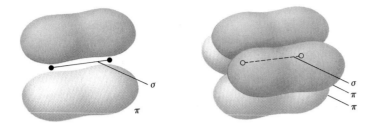

A double bond (left) consists of two electrons in a σ orbital (depicted by the line joining the nuclei) and two more electrons in a π orbital. A triple bond (right) consists of two electrons in a σ orbital and four more electrons in two π orbitals. The actual π-electron distribution is cylindrical in a triple bond (as depicted for the C_2 molecule on p. 37).

and augment the σ bonds between the atoms. Only one pair of electrons from each atom participates in the π orbital; the lone pairs are not involved in bonding and the other electron pairs are taken up by the three σ bonds. We shall proceed in two stages: first by considering the orbitals of a triangle of oxygen atoms, and then by dropping a carbon atom into their center.

The three oxygen $2p$ orbitals we consider form the three distinct superpositions shown in the illustration below. In one, there is constructive interference between all three neighbors; in the other two there is destructive interference between at least one neighboring pair. These three combinations are not yet molecular orbitals because they spread over only the three oxygen atoms, not all four atoms of the ions.

When we drop the carbon atom into the center of the ring of oxygen atoms, its perpendicular $2p$ orbital has the same symmetry as the combination of oxygen $2p$ orbitals on the left, and can overlap that combination constructively or destructively. The constructive combination (1π) is a bonding molecular orbital, and the destructive combination (2π) is antibonding. The carbon $2p$ orbital has zero net overlap with the other two combinations (there are equal amounts of constructive and destructive interference between the carbon and oxygen atoms), and so does not form a bonding or antibonding combination with them.

There are six electrons to be accommodated in the orbitals we have described (one from each atom, and two for the double negative charge of the ion). Two enter the 1π orbital: it is bonding between the carbon and oxygen atoms and between the oxygen atoms, so its presence pulls all the

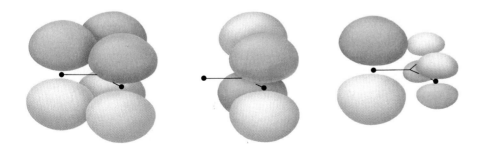

The combinations of the 2p orbitals of the three oxygen atoms in a carbonate ion, CO_3^{2-}, from which molecular orbitals will be constructed. Only the leftmost combination can overlap with a perpendicular 2p orbital on the carbon atom.

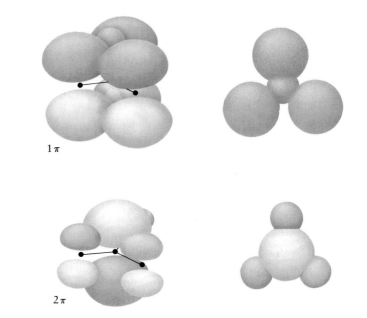

The bonding (1π) and antibonding (2π) orbitals formed by overlap of the oxygen-orbital combination and the carbon 2p orbital that lies perpendicular to the plane of the molecule. The illustrations on the right are vertical views. Note that the bonding combination is predominantly oxygen-orbital in character (as shown by the sizes of the individual orbitals), whereas the antibonding combination is predominantly carbon-orbital in character. Oxygen is more electronegative than carbon.

1π

2π

atoms together and augments the strength of the σ bonds. The remaining four occupy the two orbitals formed from oxygen-only combinations, which have no bonding character between the carbon and oxygen atoms but are slightly antibonding between neighboring oxygen atoms. Therefore, these electrons tend to weaken the carbon–oxygen bonds, but not by very much. The electrons in these orbitals tend to open out the OCO angles, but since all three angles are affected equally, the net shape of the ion remains an equilateral triangle. If electrons were to occupy the 2π orbital, they would be antibonding between the central carbon and all three oxygen atoms, and would tend to pull the ion apart symmetrically.

The only remaining detail of the structure of CO_3^{2-} that we need to note, because it is a general phenomenon, is the role of the electronegativities of the atoms. Oxygen is more highly electronegative than carbon, and so sucks electrons away from its neighbor. This can be seen clearly in the molecular orbital structure, where we see that the *bonding* orbital 1π has a much greater contribution of oxygen atomic orbitals than the central carbon atomic orbital (the antibonding orbital behaves oppositely, but that need not concern us). Wherever we have a highly electronegative element in a molecule, it makes the greater contribution to the bonding molecular orbital, and hence hogs the electron distribution in that orbital. We can contrast the structure of the carbonate ion with a fragment of soot in the flame that consists of four carbon atoms: all four atoms have the same electronegativity and the 1π molecular orbital is much more evenly spread over them.

CHAPTER TWO

I would like to imagine that Faraday would be gazing now at his flickering candle with the light of greater comprehension in his eyes. He would know that the atom of carbon in the flame was not just a solid microscopic ball of matter, but had a structure and an inner life. He would perceive the atom as a series of electrons in concentric spheres, and know that electron spin and the Pauli principle gave it bulk. He would also know something of how the atom linked to others, by blending its orbitals with theirs, and allowing their electrons to spread between them. Here too he would imagine shells of electrons, and think of the shapes of the concentric but more convoluted shells as arising from the superposition of the atomic orbitals of the parent atoms. He would know that when an electron distribution shook and discarded excess energy, that that was the source of the light from the flame.

None of this would have diminished his delight: it would have prepared his imagination to follow the atom as it left the flame.

THE MIGHT OF MINIMAL CHANGE | 3

One of the many possible destinations for a
carbon atom liberated by a flame is the
limestone that helps form our landscapes.
After many transformations of the type we
identify in this chapter, a carbon atom long
ago contributed to these fossil shells
(Promicrocevas) and has lain dormant as a
form of calcium carbonate.

You will be astonished when I tell you what this curious play of carbon amounts to.

MICHAEL FARADAY, LECTURE 6

Our aim is to pursue Faraday's liberated, flame-born carbon atom with the eyes of a modern chemist, and to see, in particular, how the atom's electronic structure—which we can now envisage—determines its future as a chemical entity. It is convenient to suppose that of all its possible fates, the atom has stumbled into combination with oxygen, and that it is carried away from the flame as carbon dioxide, CO_2. This simple triatomic molecule is the most remarkable of the combustion's products, for it is carbon's mode of entry into the lithosphere, the hydrosphere, and the atmosphere, and it is its vector for return to the biosphere.

One of the intellectual pleasures of chemistry is that it adds to our appreciation of the world by showing that what may seem to be an infinite variety of reactions can be analyzed into a tiny handful of different types. As we explore the reactions that are open to carbon, we shall see that they illuminate the various classes of reaction that chemists have identified, often by identifying the subatomic particle that one substance passes to another. Indeed, in one form or another, all the classes of chemical reaction are illustrated in some degree by carbon dioxide, so the chapter spans the whole of chemistry: just as there is a scientific world within a candle flame, so there is a boundless laboratory in a molecule. However, like Faraday, we shall sometimes shift our focus to other species when they illuminate some of the reactions more sharply than the candle.

There is also a second track that runs through this chapter, one that will take us on a journey through the conceptual structure of chemistry. Ever since the emergence of modern chemistry in the eighteenth century, chemists have sought broader views of the events that actually take place at an atomic level when a reaction occurs and matter is transformed from one form into another. In particular, they have found that one class of reaction could often be generalized—by refurbishing a definition—to incorporate what previously had been regarded as another entirely separate class of reaction. We shall capture the spirit of this enterprise in this chapter, for we shall see how the everyday concept of "acid" has been generalized by successive waves of insight to apply to an almost limitless range of substances.

THE FIRST CLASS: NEUTRALIZATION

A carbon atom travels toward an inorganic destination when the carbon dioxide that carries it away from the flame dissolves in the oceans to form carbonic acid, H_2CO_3. This acid is the parent of the carbonates, which are compounds that contain the carbonate ion, CO_3^{2-}. Carbonates occur in great abundance, for they include the calcium carbonate, $CaCO_3$, of chalk and limestone hills and the shells of crustacea, and the dolomite, $CaMg(CO_3)_2$, of the Dolomites. For our present purposes, the formation of carbonic acid opens the door to the participation of the carbon atom in the class of reactions called "neutralizations."

A neutralization is the reaction between a type of compound called an "acid"—such as carbonic acid—and another class of compound called a "base." As we shall see, chemists have put a considerable intellectual effort into deciding what they mean by the terms "acid" and "base." The definitions that we shall take as our starting point were proposed by the Swedish chemist Svante Arrhenius toward the end of the nineteenth century. Arrhenius classified a compound as acid or base according to its behavior when it is dissolved in water to form an "aqueous solution" and its molecules or ions drift apart and mingle with the water molecules. He suggested that a compound should be classified as an "acid" if it contains hydrogen and releases hydrogen ions, H^+, when it dissolves in water. The H_2CO_3 mole-

The Dolomite mountains in Austria and northern Italy are abundant supplies of dolomite, $CaMg(CO_3)_2$, one of the limestone minerals.

cule releases its hydrogen ions in water to form bicarbonate ions, HCO_3^-, and carbonate ions, CO_3^{2-},

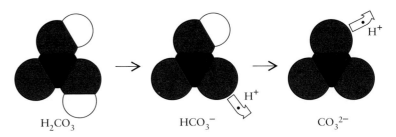

H_2CO_3 HCO_3^- CO_3^{2-}

Therefore, according to Arrhenius, H_2CO_3 is indeed an acid. We shall call any substance that behaves in the same way an "Arrhenius acid" (an A-acid) to remind ourselves that its classification as an acid depends on a particular definition. Acetic acid, CH_3COOH, the bitter component of vinegar, is another compound that can release one of its hydrogen atoms in water (the one attached to an oxygen atom), so it too is an Arrhenius acid. That carbonic acid contributes to the tingling sensation of carbonated water and that vinegar is bitter are both consequences of the hydrogen ions that the molecules release in water, for these ions stimulate sensors on the surface of the tongue.

Acetic acid, CH_3COOH

Ammonia, NH_3, dissolves in water very freely, but because it does not release any of its three hydrogen atoms to any significant extent, it is not an Arrhenius acid. In fact, ammonia is a member of the class of substances that Arrhenius defined as a "base." According to Arrhenius, a base (an "A-base") is a compound that gives rise to hydroxide ions, OH^-, in water (it need not contain OH^- ions itself: it need only generate them when it dissolves). Ammonia generates OH^- ions by the reaction

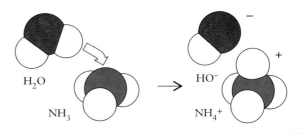

H_2O NH_3 HO^- NH_4^+

and the neutral NH_3 molecule becomes the ammonium ion, NH_4^+.

When aqueous solutions of an Arrhenius acid and an Arrhenius base react with each other in a neutralization reaction, the product is a "salt" and water:

$$\text{A-acid} + \text{A-base} \longrightarrow \text{salt} + \text{water}$$

The acid and base have mutually neutralized each other in the sense that neither the salt nor the water is an A-acid or an A-base: the acid and base character of the reactants have annihilated each other. For example, the reaction between aqueous solutions (denoted *aq*) of carbonic acid and the base sodium hydroxide, NaOH, is a solution of the salt sodium carbonate and water as a liquid (denoted *l*):

$$H_2CO_3(aq) + 2NaOH(aq) \longrightarrow Na_2CO_3(aq) + 2H_2O(l)$$

In general, a salt is an ionic compound that is the product of the neutralization of an A-acid by an A-base. Neutralization reactions can be used to obtain an almost infinite variety of salts by choosing an acid and a base appropriately; indeed, sodium hydroxide is the "base" for the formation of any sodium salt, such as sodium carbonate by reaction with carbonic acid or sodium acetate by reaction with acetic acid.

It is not difficult to identify the *essential* feature common to all Arrhenius acid-base neutralization reactions. The essential component of an acidic solution is a hydrogen ion (the acid itself is just the provider of these ions), and that of a basic solution is a hydroxide ion (provided by courtesy of the base); hence it is easy to suspect that the essential feature of an Arrhenius neutralization is the reaction of these two species:

Svante August Arrhenius (1859–1927).

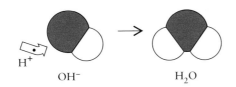

Thus, when carbonic acid is mixed with aqueous sodium hydroxide, the hydrogen ions from the acid combine with the hydroxide ions supplied by the base, and both are removed as water, leaving Na^+ and CO_3^{2-} ions in solution. Very little has happened—the original partners of the sodium ions and the carbonate ions have been plucked away, and the substance remaining is a salt. The *essential* feature of neutralization (in the Arrhenius theory) is therefore the mutual elimination of hydrogen ions and hydroxide ions, whatever the specific identity of the acid and base that supplies them in the solution.

In identifying the essential feature of a neutralization reaction we have introduced the theme that is very characteristic of chemistry and that we shall push to its limit in this chapter. If we can identify the central feature of a reaction, we may thereby discover that many disparate reactions are in fact one. The Arrhenius definitions of acids and bases are a good example of this, for they switch attention away from the entities that are added to water to the species (H^+ and OH^-) that these entities supply and which carry out the actual work of the reaction. His definitions are the starting point for the hunt for a still wider definition of an acid and a base.

BRØNSTED ACIDS AND BASES

In 1923 the Danish chemist Johannes Brønsted and the English chemist Thomas Lowry independently proposed a broader definition of acids and bases. Their proposals were motivated by the observation that the reaction between ammonia and hydrogen chloride, which in water is a typical acid-base neutralization producing the salt ammonium chloride, NH_4Cl, and water, also produces the salt when the two gases (denoted *g*) mix in the absence of water:

$$NH_3(g) + HCl(g) \longrightarrow NH_4Cl(s)$$

This reaction is a "neutralization" in the sense that one species reacts with another to give a substance that is neither an acid nor a base; however, it takes place in the absence of the solvent water, which was specifically required by the Arrhenius definitions. Moreover, advances in chemistry at the opening of the twentieth century had allowed chemists to use solvents other than water, including liquid ammonia and liquid sulfur dioxide, and it was found that analogs of acid-base neutralization in which no water was involved were very common. Arrhenius had clearly been too restrictive in his definitions.

Brønsted and Lowry identified a feature that enables us to speak of a species as *intrinsically* an acid or a base, independent of its behavior in water. What we now call a "Brønsted acid" (a B-acid) is a proton donor and a "Brønsted base" (a B-base) is a proton acceptor. (The commemoration of Lowry's contribution appears somewhat unfairly to have slipped away from common usage.) In the context of acid-base reactions, a proton is synonymous with a hydrogen ion, H^+ (a hydrogen atom consists of a proton surrounded by one electron, so the loss of the electron to form H^+ leaves only a proton).

Carbonic acid is a Brønsted acid because the H_2CO_3 molecule can donate a proton to another molecule; the ammonia molecule is a Brønsted base because it can accept a proton (such as from an HCl molecule) and form NH_4^+. Now we see that the reaction between ammonia and hydrogen chloride stated above is indeed an acid-base neutralization, since it involves the transfer of a proton from a B-acid (HCl) to a B-base (NH_3).

One strength of the Brønsted-Lowry definitions is that they extend the range of what may be considered an acid or base without overturning

When gaseous ammonia and hydrogen chloride mix, they give rise to the salt ammonium chloride, which forms as a white cloud of tiny crystals.

Johannes Nicolaus Brønsted (1879–1947).   *Thomas Martin Lowry (1874–1936).*

Arrhenius's contribution. They are *evolutionary,* not *revolutionary* redefinitions. The theory shifts our focus of attention away from aqueous solutions, so that the concept of acid and base applies whatever the solvent and indeed applies even in the absence of a solvent. It also shifts our attention on to the fundamental act involved in the reactions of acids and bases, the transfer of an elementary particle—the proton—from one species to another. All the familiar reactions that depend on the action of acids and bases can be understood in terms of this simple act. The theory brings the proton into the heart of chemistry, and in particular allows us to comprehend a great deal of chemistry as the consequences of the transfer of a proton.

However, we should pause at this point to ask whether we really want the proton to play such a dominant role! Chemistry, we remarked in Chapter 2, concerns the behavior of *electrons,* not protons, so should we be happy with this sudden shift of focus? A more practical point is that even the Brønsted-Lowry theory does not capture all substances that behave in some respects like acids and bases. For example, the reaction between carbon dioxide and calcium oxide (CaO, quicklime)

$$CaO(s) + CO_2(g) \longrightarrow CaCO_3(s)$$

is a kind of neutralization reaction in which two compounds satisfy each other's tendency to react, just like the reaction between an acid and a base. However, there is no proton transfer in the reaction, so it is not a Brønsted-

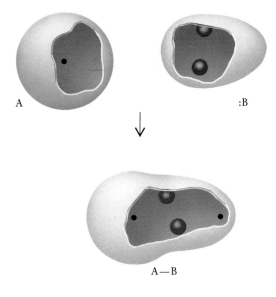

The essential feature of a reaction between a Lewis acid and a Lewis base is the formation of a covalent bond when one species (the base) donates a pair of electrons to the other (the acid).

Lowry neutralization. Could it be that there is an even broader theory of acids and bases that concentrates on electrons and does not depend on the presence of a transferrable proton?

LEWIS ACIDS AND BASES

Gilbert Lewis, whose preoccupation with electron pairs in 1916 led him to formulate a powerful theory of molecular structure (as we saw in Chapter 2), harvested further fruits of his concept when in 1923 he formulated a very broad definition of acid-base behavior. Lewis considered that the crucial attribute of an acid is that it can accept a pair of electrons, and the crucial attribute of a base is that it can donate a pair of electrons. In the context of this theory, donation results in the formation of a covalent bond between the acid and the base; it is not complete transfer. In his definitions, Lewis switches attention away from the proton and refocuses it on an electron pair. However, we shall see that in doing so, his definition encompasses Brønsted acids and bases, and leaves the splendid achievements of the Brønsted-Lowry theory intact. His redefinitions are also evolutionary, not revolutionary.

First, consider a Lewis base (an "L-base"), a species that can donate an electron pair and thereby form a covalent bond. An oxide ion, O^{2-}, rates as a Lewis base because it has a lone pair of electrons that it can donate to a suitable acceptor. Any species that acts as a B-base is also an L-base because a species that can accept a proton must possess a pair of electrons to which the proton can attach. When we are dealing with a species that uses one of its lone pairs to act as a Lewis base, it is convenient to display that pair explicitly; so from now on we write the oxide ion as $:O^{2-}$. If a species has several lone pairs (the oxide ion has four), we often show only one of them unless all of them are important. The hydroxide ion is another L-base, and we represent it as $:OH^-$ (the ion in fact has three lone pairs, $:\overset{..}{O}H^-$).

Now consider a Lewis acid (an "L-acid"), which is a species that can accept a pair of electrons and thereby form a covalent bond with the donor (the Lewis base). A very important example of an L-acid is the proton itself. Thus, a proton, H^+, can attach to an electron pair provided by a Lewis base, and form a covalent bond with it:

$$H^+ + :O^{2-} \longrightarrow O—H^-$$

The class of L-acids encompasses all the B-acids, but there is a slight twist. A B-acid *contains* a proton, so it acts as a *source* of the L-acid. Thus carbonic acid, which is both an A-acid and a B-acid, is the source of an L-acid.

It is of the greatest relevance to our story that carbon dioxide is a Lewis acid; it is certainly not a Brønsted acid because it has no protons to release. Its action as an L-acid can be seen when carbon dioxide molecules dissolve in water, as in the carbonation of a drink or, on a more global scale, when the gas dissolves in the oceans. The carbon dioxide reacts with hydroxide ions that are naturally present in water from the ionization of a small proportion of the H_2O molecules:

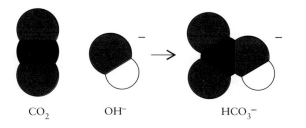

$$CO_2 \qquad OH^- \qquad \rightarrow \qquad HCO_3^-$$

Portland cement is made by heating a mixture of silica, clay, and limestone to about 1500°C. The cooled mass is then ground to a fine powder and some gypsum ($CaSO_4 \cdot 2H_2O$) is added. Reaction occurs when water is added to the mixture, and it sets into a solid mass consisting largely of calcium silicate. The interlocking crystals that give cement its strength can be seen in this micrograph.

A lone pair of the hydroxide ion (a Lewis base) insinuates itself into the orbitals of the carbon atom, driving back the electrons of one of the C=O double bonds on to the oxygen atom and forming a new C—O link: the result is the HCO_3^- ion, which is only a proton away from being carbonic acid. The carbon atom of the CO_2 molecule has accepted a lone pair of electrons in the reaction, so carbon dioxide is a Lewis acid.

An almost identical process accounts for the reaction that ended the last section, the formation of calcium carbonate when calcium oxide (quicklime) is exposed to carbon dioxide, but in this reaction the CO_2 molecule is speared by the lone pair of an oxide ion. A very similar reaction occurs when mortar—a mixture of slaked lime, $Ca(OH)_2$, and silica, SiO_2, in the form of sand—sets on exposure to air to form a mass of calcium carbonate and silica. The reverse of the carbon-dioxide spearing reaction—the thermal decomposition of limestone—is used to make cement and as a source of lime in the blast furnace for ironmaking. We can now easily comprehend the processes that create quicklime from limestone in a hot furnace, for as the carbonate ions vibrate very vigorously, the oxide ion may shake off the Lewis acid CO_2 that was bonded to it:

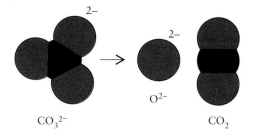

$$CO_3^{2-} \qquad \rightarrow \qquad O^{2-} \qquad CO_2$$

The calcium ions trap the oxide ions in the solid and the carbon dioxide molecules fly off as a gas.

The formation of carbonate ions when carbon dioxide dissolves in water or is captured by an oxide ion is a special case of a fundamental type of reaction between a Lewis acid and Lewis base. This reaction is the formation of a "complex," in which the L-acid and the L-base become bound together by a covalent bond that is formed by the electron pair donated by the base to the acid:

$$A \ + \ :B \ \longrightarrow \ A—B$$

L-acid   L-base   Complex

This fundamental reaction type (and its reverse, the decomposition of a complex) summarizes a wide range of reactions and is the gateway through which carbon dioxide passes into the ocean and the landscape.

It turns out that many chemical reactions involve a slight modification of the fundamental complex formation reaction. One modification is the "displacement reaction," in which an L-base (:B′) drives out another (:B) from a complex:

$$B—A \ + \ :B′ \ \longrightarrow \ B: \ + \ A—B′$$

Complex   L-base   L-base   Complex

This displacement reaction is the pattern of a large number of reactions, including all Brønsted acid-base reactions. A variation of the same theme is that one L-acid can drive out another:

$$A′ \ + \ B—A \ \longrightarrow \ A′—B \ + \ A$$

L-acid   Complex   Complex   L-acid

An example of a displacement reaction is the one that Faraday actually used in his lecture to produce carbon dioxide. He added chips of "a very beautiful and superior marble" (calcium carbonate) to a flask containing "muriatic acid" (hydrochloric acid, an aqueous solution of hydrogen chloride, HCl). We know today that the essential feature of the reaction

$$CaCO_3(s) + 2HCl(aq) \longrightarrow CaCl_2(aq) + H_2O(l) + CO_2(g)$$

The action of dilute hydrochloric acid on chips of marble can be used to produce carbon dioxide in the laboratory.

is the decomposition of the carbonate ions in the solid by the action of the ever-pernicious hydrogen ions supplied by the acid. Neither the calcium cations (which merely pin the carbonate ions together in the solid) nor the

CHAPTER THREE

chloride ions (which accompany the hydrogen ions in the acid) play any direct role in the reaction (they are "spectator ions"); if they and one of the protons are removed from both sides of the equation, we obtain:

$$CO_3{}^{2-}(s) + H^+(aq) \longrightarrow CO_2(g) + OH^-(aq)$$

With the skeleton of the reaction exposed we can identify its type. We have already seen that the carbonate ion is a complex of the L-acid CO_2 and the L-base $:O^{2-}$. Moreover, the OH^- ion is a complex of the L-acid H^+ and the L-base $:O^{2-}$. Hence, the reaction can be pictured as a displacement in which the H^+, a very powerful L-acid, drives out the weaker CO_2 L-acid from its complex with the O^{2-} ion:

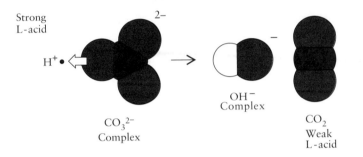

In his simple experiment, Faraday was unknowingly conjuring with lone pairs and inducing a base to leave its weaker partner.

An example of displacement by a second L-base—which gives an indication of why displacement reactions are ubiquitous—occurs between carbonic acid and the water that surrounds it in a raindrop or an ocean. In that reaction, a proton (an L-acid) is transferred from an acid (a complex) to water, which can act as an L-base because it has a lone pair available:

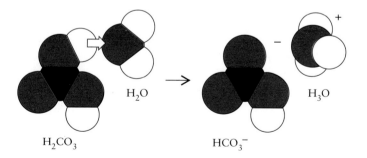

Similarly, the formation of H_2CO_3 when $HCO_3{}^-$ ions are added to water is also a displacement reaction, but now water plays the part of the complex

H—OH. In this case, the HCO_3^- ion is the incoming L-base that plucks a proton (an L-acid) out of its combination with a hydroxide ion (another L-base):

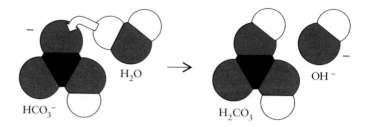

Now that we have come this far with recreating the fundamental processes of the Brønsted-Lowry theory of acids and bases, it should not be surprising that we can recreate the process that underlies neutralization in water:

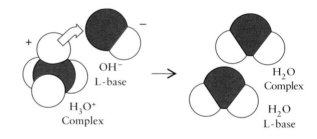

In brief, the fundamental process of Brønsted-Lowry theory—proton transfer—is a displacement reaction in the Lewis formalism. We should note the chameleonlike character of water: H_2O acts as both an L-base and a complex in the same reaction. It is also a B-acid (because it can donate a proton to another species) and a B-base (because it can accept a proton and become a H_3O^+ ion). We are beginning to discover water's unique ability to act in a wide variety of ways in chemical reactions.

There is an important shift of emphasis between the Brønsted-Lowry and Lewis theories of acids, just as there was between the Arrhenius and Brønsted-Lowry theories. A B-acid is the entity HA, the proton donor. In the Lewis view of the entity HA, the L-acid is the H^+ ion (the proton) itself, and the B-acid is simply the source of this acid. The central difference between the two approaches is that the Lewis theory concentrates on the actual entity that is central to the behavior of the species. We shall pursue the spirit of this intellectual adjustment again later in the chapter, where we push it to (some might think, through) its logical conclusion.

When a solution of sodium sulfide is poured into a solution of calcium nitrate, a cloudy yellow precipitate of calcium sulfide is formed.

CHAPTER THREE

THE SECOND CLASS: PRECIPITATION

Suppose our carbon atom is present in a carbonate ion in a solution of sodium carbonate, Na_2CO_3, and that we also have a solution of the very soluble compound calcium chloride, $CaCl_2$. The two compounds are soluble in water largely because both the sodium cations (Na^+) and chloride anions (Cl^-) are quite big and carry only a single charge; consequently, they do not attract their respective partners (the carbonate anions and the calcium cations) very strongly, and when the two compounds are placed in water, all the ions separate and disperse through the liquid. When the two solutions are mixed, sodium and chloride ions remain in solution, but a dense white "precipitate"—an insoluble solid that seems to fall out of solu-

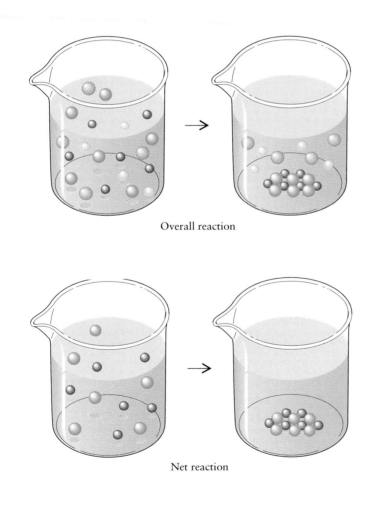

Overall reaction

Net reaction

The essential feature of a precipitation reaction (top) can be clarified by removing the "spectator ions," the ions that remain unchanged in solution (bottom). This shows that in the reaction the only change is that two types of ion leave the solution by forming an insoluble solid.

tion—immediately forms. The precipitate consists of the calcium carbonate, and the reaction that produces it is

$$Na_2CO_3(aq) + CaCl_2(aq) \longrightarrow CaCO_3(s) + 2NaCl(aq)$$

| Sodium carbonate solution | Calcium chloride solution | Calcium carbonate precipitate | Sodium chloride solution |

A new substance—calcium carbonate—has been created out of two other substances, sodium carbonate and calcium chloride. Happily for our landscapes, buildings, and sculptures, calcium carbonate is not very soluble in water, largely because the small, highly charged Ca^{2+} ion has a strong attraction for the highly charged CO_3^{2-} ion. The ability of the Ca^{2+} ion to

A stalactite is a deposit of calcium carbonate that hangs from the roof of a limestone cavern. The name is derived from the Greek words meaning to fall in drops. A stalagmite ("that which drops") is the corresponding columnar structure that rises upward from the floor.

CHAPTER THREE

attract other ions strongly is at the root of the reasons why calcium is so widely used as a structural material: as a carbonate it is found as limestone and marble, as a phosphate it is a component of our bones, and as a silicate it is a component of cement.

Despite all the cloudy turbulence that accompanies the reaction, nothing very much has changed, for all the original ions are still present. The calcium cations and carbonate anions that are suddenly found in the same solution have clumped together and have formed an insoluble solid, leaving the accompanying ions in the solution. The reaction—almost a nonreaction, since there has been the merest exchange of partners—is called a "precipitation reaction." In general, the net outcome of a precipitation reaction is the falling out of solution of two types of ions on account of the insolubility of the compound they jointly represent.

The impression must not be gained that all our limestone hills and mountains are the result of some gigantic precipitation reaction. Although some limestone deposits have formed by direct precipitation (for instance, where upwelling streams of calcium-laden water enter oceans), most have formed indirectly as marine life has deployed calcium and carbonate ions to grow shells and then has discarded these temporary habitations in death. The actual process of mobilization of calcium for shell formation, coral growth, and the formation of our bones and teeth is under biological control, and when the system fails it may lead to the deposit of calcium carbonate (and phosphate) in inappropriate locations, resulting in stone formation, cataracts, and osteoarthritis.

THE THIRD CLASS: RADICAL REACTIONS

When a covalent bond is formed or broken in a Lewis acid-base reaction, one species supplies a bond's entire electron pair. In the next type of reaction we consider, each species supplies *one* electron to a bonding pair. A species with an unpaired electron is called a "radical," and a typical reaction is one in which two radicals link by forming a covalent bond. An example is the combination of a hydroxyl radical, ·OH (the single dot denotes the unpaired electron), and a methyl radical, CH_3·,

$$CH_3 \cdot \; + \; \cdot OH \longrightarrow CH_3OH$$

which is one of the enormous number of reactions that take place in a flame as the fuel is ripped apart on its route to forming carbon dioxide.

The combination of two radicals is the one-electron analog of a Lewis acid-base neutralization, in which an "acid" (actually a radical) and a

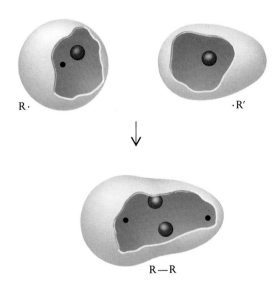

The essential feature of a radical combination reaction is the formation of a covalent bond when each radical provides one electron.

"base" (another radical) combine to form a "complex" (a molecule with all its electrons paired) that is neither an "acid" nor a "base." Indeed, the analogy extends beyond "complex" formation, because there is also a type of reaction between radicals that is the analog of a displacement reaction. In such a reaction, a radical attacks a molecule and drives out another radical, just as in a displacement reaction an L-acid attacks a complex and drives out another L-acid. For example, a hydroxyl radical (·OH) can attack a chlorofluorocarbon molecule that contains a C—H bond, leading to the formation of a new radical:

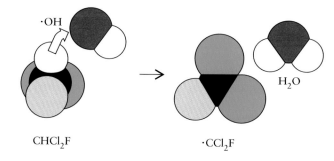

CHCl₂F   ·CCl₂F

The form of this reaction,

$$\text{Radical} + \text{molecule} \longrightarrow \text{molecule} + \text{radical}$$

is the analog of

$$\text{L-acid} + \text{complex} \longrightarrow \text{complex} + \text{L-acid}$$

Incidentally, the reaction given above has given some hope of a reprieve for chlorofluorocarbons. The current generation of chlorofluorocarbons are typically species such as CF_2Cl_2 that have had all their hydrogen atoms replaced by halogen atoms. Having no tender C—H bonds to make them vulnerable to destruction, they ascend through the troposphere and into the stratosphere, where they are fragmented into radicals by the hammer blow of solar radiation. These radicals react with ozone, O_3, lower its concentration, and so expose the life below to harmful radiation. However, if a C—H bond is present, the chlorofluorocarbons are open to attack by hydroxyl and other radicals while they are still in the troposphere, and hence might never reach the stratosphere. The possibility thus arises of producing chlorofluorocarbons that pose no threat to the ozone layer. Another very significant reaction in the troposphere is the reaction that helps to eliminate the carbon monoxide that is present in the air:

$$HO\cdot + CO \longrightarrow H\cdot + CO_2$$

This reaction plays a dominant role in governing the chemical composition of the atmospheres of Venus and Mars, on which carbon monoxide is formed by the effect of solar radiation on carbon dioxide, which is highly abundant in their atmospheres.

Radical reactions also play a role in speeding the recycling of organic forms of carbon, for they have been implicated in degenerative diseases, senescence, and death. For example, by reactions akin to burning, radicals may attack the hydrocarbon-like molecules of cell membranes, causing cell death. Indeed, new forms of therapy are being directed specifically to counter radical reactions. For instance, compounds such as vitamins E and C are known to trap oxygen radicals before they can do substantial damage to cell membranes; current research is exploring the same strategy to combat neurological damage caused by oxygen-containing radicals that form after head and spinal cord injuries and stroke. The radicals form when reactions go awry and release intermediates before they have been fully converted into products. In a typical radical-damming strategy, the virulent radical, which might be a peroxyl radical, $\cdot$OOH, attacks the protecting species and either links to it, in which case the radical is prevented from doing further damage, or forms a new radical that is too big to move from the site of its formation.

Radical reactions are essential to the combustion of fuels, and when Faraday lit his candle he was releasing a Niagara of radicals. The act of ignition—with a glowing filament or another flame—decomposes some hydrocarbon molecules and generates radicals, as in the reaction

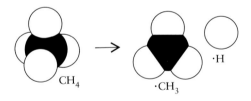

These radicals then launch an attack on other species present, as in

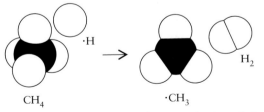

Once some radicals have been formed, the complexity of the reactions that may ensue is almost unlimited. Thus, more hydrogen atoms can be plucked off the methyl radicals, and the resulting carbon-containing radicals can attach to each other and grow into particles of nearly pure carbon, so pro-

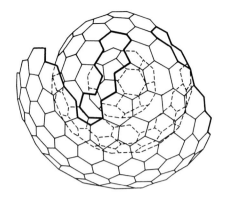

There is some evidence that the particles of carbon that initially form in the smoke of a flame have the form of spherical coils of chicken wire.

ducing smoke. Should the burning mixture be poor in oxygen, as in a candle flame, the groups of carbon atoms may grow until each particle consists of thousands of atoms. These particles, at the temperature of the flame, are incandescent: their excited electrons emit radiation of a wide range of wavelengths and glow yellowish white.

Under some conditions of temperature and pressure (but never in a candle), a mixture of hydrocarbons and air may explode. We can identify one process leading to explosion by noting that some steps of the reaction increase the number of radicals. For example, the step

$$H\cdot\ +\ O_2\ \longrightarrow\ HO\cdot\ +\ \cdot O\cdot$$

generates two radicals from one on the left (an oxygen atom has two unpaired electrons). The technical term for a reaction step in which the number of radicals increases is "chain branching." As a result of chain branching, with each radical able to initiate its own progeny of reactive radicals, the rate of the reaction may increase dramatically and the fuel may be consumed explosively quickly—the fuel-air mixture may explode.

Once we know that combustion is propagated by radicals, we can identify a way to control unwanted conflagration. The combustion will die away if we can intrude into the network of reactions and divert or terminate the radical chain. One procedure is to inject halogen atoms into the flame, for they can extract a hydrogen atom from a hydrocarbon molecule, as in the reaction

$$Br\cdot\ +\ H{-}CH_3\ \longrightarrow\ Br{-}H\ +\ \cdot CH_3$$

Now the HBr molecule is a lure for attack by a hydroxyl radical,

$$HO\cdot\ +\ H{-}Br\ \longrightarrow\ HO{-}H\ +\ Br\cdot$$

The radical chain is diverted from the hydrocarbon, and the halogen atom is regenerated and can reenter the cycle of diverting reactions. This radical-distracting process is the rationale behind some types of fire extinguisher. Halogen-containing compounds are also embedded into or actually combined with fabrics to make them flame-retardant, for as the fabric begins to burn, the compounds release bromine atoms and terminate the chain.

The combustion of a solid fuel, like candlewax and carbohydrate in the form of wood, differs from that of vapors and gases inasmuch as that before it can burn it must first be partially decomposed and vaporized. Thus, the initial step in the combustion of a solid is the fragmentation of molecules close to the surface, and their escape into the space above, where they burn by the radical process that we have described. This surface fragmentation is sustained by the combustion taking place above it. The layer of flame close to the surface supplies heat which batters the surface with such intense radiation that bonds are broken and the fragments are ejected from the surface before they can recombine.

The knowledge that surface fragmentation is an essential step in the propagation of a flame gives another route to the elimination of conflagration: it may be possible to include materials in a surface that inhibit its decomposition and fragmentation. If phosphates are incorporated, for example, the initial heat of the flame will set off reactions that help to seal the surface and prevent its fragmentation into vaporized fuel.

As well as producing heat, radical reactions can be harnessed for work. In contrast to heat, which is energy released as the chaotic motion of atoms, work is energy released as the uniform movement of atoms. The paradigm of work—the process to which all work is at least hypothetically equivalent—is the lifting of all the atoms of a weight to a greater height. Indeed, the delay between the discovery of the use of fuel to heat and the use of fuel to drive reflects the much greater technological challenge of finding a way to release motion in a coordinated way rather than in the disorderly muddle of heat.

To achieve work, a reaction is carried out inside an *asymmetrical* system. In an internal combustion engine, the asymmetrical system is a cylinder equipped with one movable wall (the piston); in a rocket engine it is a cylinder with a simple orifice (or a not so simple orifice in a jet engine). The asymmetry of the reaction cavity allows us to extract the energy of combustion as motion. The heat released by the reaction expands the gases within the cavity, and the greatly increased temperature causes the product molecules, radicals, and ions to travel at much greater speeds. Their impact on the movable wall drives it outward, and it is a straighforward engineering problem to convert this linear motion into rotational motion (and, in many cases, back into linear motion again as the car is driven along the road). If there is no wall, as in a rocket, the momentum of the departing molecules impels the vehicle forward.

In a gasoline engine, the fuel should burn smoothly through the power stroke to give a uniform thrust against the moving piston. In contrast, in a diesel engine, the fuel burns essentially explosively when the gases are fully compressed. Therefore, we need two types of combustion: a lengthy burn for the gasoline engine and a rapid combustion for the diesel engine. It is found that straight-chain hydrocarbons burn rapidly: their carbon chains are so exposed that they can readily shatter into radicals, and the combustion reaction is explosively fast. Hence, by using a fuel that is rich in straight-chain hydrocarbons, we can meet the requirement of a diesel engine for a sudden release of energy at the top of its compression stroke. Typical

Iso-octane, C_8H_{18}

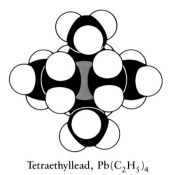

Tetraethyllead, $Pb(C_2H_5)_4$

diesel fuels are based on hydrocarbons such as cetane (hexadecane), which consists of molecules with 16 carbon atoms in a linear string. On the other hand, the substance known as iso-octane has a more protected structure, and it participates in radical reactions much less explosively: it burns, but it burns smoothly. The same is true of compounds related to benzene, such as toluene and xylene, where the benzene ring confers an intrinsic stability on the molecules, and they are protected against overly fast reaction.

Part of the work of a refinery is to conjure from crude oil the molecules with the appropriate combustion characteristics. An alternative procedure is to add to the fuel a substance that participates in the reaction and modifies its course. The infamous compound tetraethyllead for long fulfilled this role, for the ethyl groups, C_2H_5, are only weakly anchored to the central lead atom, and are blown off as ethyl radicals when the combustion reaction is initiated. They can then participate in the radical chain reaction that corresponds to the combustion reaction, and hence contribute to its smooth propagation.

THE FOURTH CLASS: REDOX REACTIONS

The pivot of Faraday's lectures, the combustion of candlewax, was a "redox reaction." Redox reactions occur in enormous variety. They include the reactions at the foundations of civilization, such as the extraction of metals from their ores. They also include the reactions at the root of biological activity, for redox reactions are the key reactions in photosynthesis and respiration. Redox reactions also encompass the processes that unconsciously seek to conquer civilization, for they include corrosion and the uncontrolled combustion of conflagration. Almost wherever we look, and in places we cannot see, a redox reaction is taking place.

We can trace the recognition of the existence of this class of reactions back to some ancient chemical observations. In the early days of chemistry, the combination of a substance with oxygen was called "oxidation." Indeed, the formation of carbon dioxide in a flame fueled by a hydrocarbon is the result of the oxidation of the hydrocarbon. Combination with oxygen is still called oxidation, but the term is now applied to a much wider class of reactions that includes combination with oxygen as a special case. Note, once again, how the mode of thought that is so typical of chemistry (and of science in general) is applied to oxidation as it was to the concept of acid and base. The original concept of oxidation was like a harpoon that could spear only a few reactions, and the modern concept is like a net that can capture a multitude.

CHAPTER THREE

We can identify the quintessence of oxidation by stepping away for a moment from the reactions of carbon and considering a reaction, the burning of magnesium in oxygen, that shows the essential feature of this type of reaction in a clearer form. The combustion of magnesium will be familiar to everyone who has seen a fireworks display, because the bright sparks are typically white-hot grains of magnesium that are undergoing the reaction

$$2Mg(s) + O_2(g) \longrightarrow 2MgO(s)$$

Magnesium burns vigorously because it reacts with the nitrogen and even the carbon dioxide of the air as well as the oxygen, but we can ignore these complications.

Redox reactions are responsible for many important processes, including conflagrations (right) and corrosion as on this part of the Titanic (left).

The oxidation of magnesium metal to its oxide is responsible for much of the brilliance of a firework display. Magnesium gives rise to the bright white light (left), yellows arise from sodium compounds, reds from strontium compounds, and blue from copper chloride molecules (right).

We can identify the essentials of the reaction by noting that, like almost all oxides, magnesium oxide, MgO, is a solid composed of cations (Mg^{2+} in this case) and O^{2-} anions. Therefore, there has been *electron transfer* from magnesium to oxygen. In particular, each magnesium atom of the metallic magnesium has lost two of its electrons:

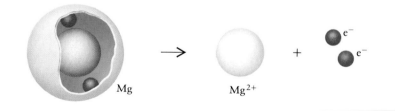

So, the essential step in the oxidation of magnesium appears to be the loss of electrons. From now on we shall elevate that observation to a definition: *oxidation is electron loss*. Faraday would have been delighted at this identification: that arch conjuror with electrons did not even know of their existence (they were discovered thirty years after his death), yet he suspected and explored the deep relationship between electricity—the motion of electrons—and the composition and reactions of matter.

We can appreciate the power of this generalization of the meaning of oxidation by considering a similar reaction in which there is no oxygen. For instance, magnesium burns in the gas chlorine, with the formation of solid magnesium chloride:

$$Mg(s) + Cl_2(g) \longrightarrow MgCl_2(s)$$

The same essential reaction has taken place: magnesium atoms have surrendered electrons to another element. Since the magnesium has undergone electron loss, according to the definition of oxidation, magnesium has been oxidized even though no oxygen is present. We are discerning the essential process that occurs during the reaction, not the plumage that accompanies it. Any species that removes electrons from another species—the roles of oxygen and chlorine in these examples—and hence brings about that species' oxidation, is called an "oxidizing agent."

Now we turn to the other wing of the same reaction. The extraction of metals from their ores (which are typically oxides, such as hematite, Fe_2O_3, a principal iron ore) is called the "reduction" of the ore to the metal. The reduction of iron ore is typically carried out with carbon monoxide, which extracts the oxygen from the ore and carries it away as carbon dioxide. In hematite, the metallic element is present as the iron ion Fe^{3+}, whereas in the metal itself it is present as electrically neutral metal atoms, Fe. The essential step in a reduction of an ore to the metal is therefore the attachment of electrons to iron cations, which we can symbolize as

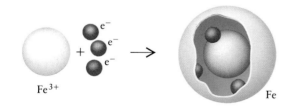

Fe^{3+} Fe

The obvious next step is to generalize by classifying as a reduction *any* reaction that involves the transfer of electrons to another species. That is, *reduction is electron gain.* (Later we shall see that atoms sometimes accompany the transfer of electrons; nevertheless, the crucial aspect of the process is the addition of electrons to the species.)

We have already met several examples of reductions. For example, in the oxidation of magnesium by chlorine, the chlorine gained the electrons that the magnesium gave up; therefore, the chlorine was reduced to chloride ions:

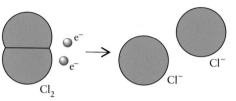

Similarly, when magnesium was oxidized by oxygen, the oxygen gained the electrons that the magnesium gave up, so the oxygen was reduced to oxide ions:

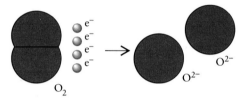

Any species that supplies electrons to another species and hence effects its reduction is called a "reducing agent."

It should be clear that, because electrons cannot simply get lost, any reaction in which one species is reduced must involve the oxidation of another species. That is, oxidation and reduction always occur jointly, with the electrons released by one species being taken up by another. This twinning of oxidation and reduction has inspired the name "redox reaction" for reactions in which electrons are transferred.

We can summarize a redox reaction as

Reducing agent + oxidizing agent → oxidized species + reduced species

A scheme such as this should be familiar: there is a hint of a similarity here between redox reactions and acid-base behavior. Whereas in the Brønsted theory the proton (one fundamental particle) is transferred from a donor (the acid) to an acceptor (the base), in a redox reaction an electron (another fundamental particle) is transferred from a donor (the reducing agent) to an acceptor (the oxidizing agent). We appear to be on the track of ascribing great areas of chemistry to the transfers of one of two fundamental particles. Can such minimal transfers be the source of all change?

The transfer of an electron can have enormous consequences. We shall see later that some of the central steps in photosynthesis involve levering an electron from one species to another. Another example of the might of such minimal change is the launching of a rocket: the transfer of an electron may blast a rocket into orbit, for rocket fuels release their energy by redox reactions of various kinds. The main engines of the space shuttle and other major rockets (such as the huge Soviet rockets and the Ariane rockets of the European Space Agency) are fueled by liquid hydrogen and oxygen, which combine in a redox reaction to give water. The tumultuous ascent of these rockets is achieved largely by hurling vast numbers of water molecules from the rear of the vehicle, and allowing it to rise by the conservation of linear momentum.

Big rockets, however, cannot rise on water alone, but require the additional thrust of disposable booster rockets. The space shuttle uses solid-fuel booster rockets that make use of the oxidation of aluminum (a light but reactive metal) by reaction with ammonium perchlorate, NH_4ClO_4. The reaction taking place within the turbulent conditions of the flaming exhaust is complex, but is broadly

$$10Al(s) + 6NH_4ClO_4(s) \longrightarrow 5Al_2O_3(s) + 3N_2(g) + 6HCl(g) + 9H_2O(g)$$

and the witches' brew of high-velocity molecules and their fragments hurtles from the rocket's flaming tail and impels the rocket upward.

The roar of a rocket is also a demonstration of the ability of a redox reaction to generate sound. In its case, the turbulence of the rocket's exhaust generates the modulating pressure waves to which our ears respond. However, the delight that reactions can conjure for some of our senses has not yet extended beyond these crude, noneuphonious rumbles and bangs. Perhaps one day a gaseous reaction that oscillates (in the sense we describe in Chapter 6) will be found that generates a harmony; for the time being, though, we must be content with bangs.

A bang is essentially a suddenly generated pressure wave. If the compression is great enough, the air behaves like a solid wall of steel flashing outward at the speed of sound, smashing and crushing whatever stands in its path. To achieve this destructive radiating wall of compression, the chemist—now bomber—must initiate a reaction (usually a redox reaction) that produces a large number of gaseous molecules in a small region. If a dozen tiny molecules are generated in the space occupied by one large molecule, the reaction has effectively produced a very highly compressed gas that will spring outward from its source.

Among the most notorious explosive substances that chemists have made available to miners, quarriers, and killers is nitroglycerin, $C_3H_5N_3O_9$. If we suppose that combustion is complete, the reaction that this substance undergoes when detonated is

$$4C_3H_5N_3O_9(l) \longrightarrow 6N_2(g) + O_2(g) + 12CO_2(g) + 10H_2O(g)$$

A rocket, here an Ariane rocket of the European Space Agency carrying a commercial payload, is impelled upward by a strongly exothermic redox reaction that results in the very hot exhaust gases rushing from the rocket engines at high speed, and hence high momentum.

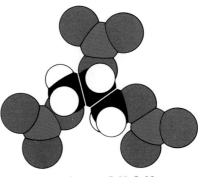

Nitroglycerin, $C_3H_5O_9N_3$

and we see that 29 small, gaseous molecules are formed, which thunder off into the night. However, such is the turmoil of the reaction, that the initial products are all manner of gaseous molecular fragments, and the final products are achieved at leisure later in the reaction. Hence, the result of detonation is better written

$$C_3H_5N_3O_9(l) \longrightarrow \text{fragments (CO, CO}_2, \text{N}_2, \text{NO, OH, . . .)}$$

and there are almost countless fragments. A point to note is that no oxygen need be supplied from the outside, and the reaction is largely a rapid rearrangement of the atoms of the original nitroglycerin molecule.

That nitroglycerin explodes on impact was immortalized in the film *The Wages of Fear*. A mechanical shock can induce some molecules to vibrate, and the swing of the atoms may precipitate the fragmentation rearrangement. The shock of that initial local expansion will stimulate fragmentation throughout the sample, and all the molecules will undergo sudden, explosive fragmentation. The sensitivity of nitroglycerin to mechanical impact is reduced by distributing the molecules through an absorbent medium. That its treacherous sensitivity could be reduced in this way was the discovery of Alfred Nobel, and contributed (together with the income from the mines he owned in Russia) to his fortune. Nobel's "Dynamite" consisted of nitroglycerin absorbed in the absorbent clay kieselguhr. The modern versions of dynamite have replaced kieselguhr with absorbent fabrics, wood flour, and paper, and the oxygen supply is enhanced (to cut down the amount of smoke by ensuring more complete combustion) by the inclusion of the oxidizing agents sodium nitrate or ammonium nitrate.

THE DETECTION OF OXIDATION

Although electron transfer is easy to identify in many redox reactions, we sometimes need to look for it with a carefully prepared eye. For example, in what sense was Faraday bringing about electron loss when he ignited the wax of his candle? What are the transfers of electrons involved in the oxidation of a hydrocarbon, or in the oxidation of carbon to carbon monoxide or carbon dioxide?

To identify the transfer of electrons in these more obscure cases, we make use of the concept of the electronegativity that was mentioned in Chapter 2. In particular, we must bear in mind that the small size of the oxygen atom makes it hungry for electrons and causes it to exert a powerful electron-withdrawing influence. Although the CO_2 molecule is not ionic, the oxygen atoms in the molecule can at least *partially* suck electrons off the central carbon atom, so *formally* it has gained electrons at the expense of the carbon atom. That the carbon atom has lost electrons (at least partially) qualifies it for classification as a species that has been oxidized. That the oxygen has partially gained electrons allows it to be classified as a species that has been reduced.

The conceptual exaggeration of these partial relocations of electrons into full-blown electron transfers enables us to apply the electron-transfer definitions to any species, ionic or covalent. For example, the reaction by which 20 million tons of nitrogen are plucked from the skies each year and converted into ammonia in its first stages of becoming food, pharmaceutical, or polymer,

$$N_2(g) + 3H_2(g) \longrightarrow 2NH_3(g)$$

is a redox reaction in which hydrogen is oxidized by nitrogen (and, equivalently, in which nitrogen is reduced by hydrogen). That it is a redox reaction can be verified by noting that the electron pairs in the N—H bonds lie closer to the electronegative nitrogen atom. If we exaggerate the tiny tendency of electrons to favor locations close to the nitrogen atom, an "exaggerated ionic" representation of the molecule would be $(N^{3-})(H^+)_3$, and the reaction has transferred electrons from the hydrogen atoms to the nitrogen.

A FURTHER SYNTHESIS

So far, we have identified four classes of reaction:

- Acid-base reactions (in Lewis terms, which subsumes Arrhenius and Brønsted behavior)
- Precipitation reactions
- Radical reactions
- Redox reactions

Are the classes distinct, or is one class of reaction another in disguise? Could it be that there is actually only *one* type of reaction?

That there is a quarry to hunt becomes apparent as soon as we realize that precipitation reactions are a special case of Lewis acid-base reactions. To see this, we need to follow the drift in the position of the electron pair in a covalent bond as we imagine increasing the electronegativity of one of the

When a solution of a soluble chloride (sodium chloride here) is poured into a solution of a soluble silver salt (silver nitrate), an immediate precipitate of silver chloride is formed.

atoms. As the electronegativity of the atom increases, the shared electron pair drifts ever closer to it, and when the electronegativity is very high, the electron pair is entirely on that atom: the more electronegative atom has become an anion and the less electronegative atom has become a cation. In other words, an ionic bond is the extreme case of a covalent bond between two dissimilar atoms.

With that thought in mind, consider what happens in a typical precipitation reaction. For simplicity, we consider the precipitation of silver chloride in the net reaction

$$Ag^+(aq) + Cl^-(aq) \longrightarrow AgCl(s)$$

which occurs when a solution of a chloride is added to a solution of a soluble silver salt. The Ag^+ ion is a typical Lewis acid: it has an empty valence shell that results in a positive charge, and it can readily attach to lone pairs. The Cl^- ion is a typical Lewis base, since it bristles with four lone pairs of electrons. The solid is a compound of the two formed when the Ag^+ ion attaches to a lone pair from a Cl^- ion. If we supposed that AgCl is purely covalent, we would have no hesitation in saying that the reaction is a Lewis acid-base complex formation. Even if we supposed that AgCl is purely ionic, we could claim that the reaction is a Lewis acid-base reaction, the only slight difference being that the "ionic" bonding is merely the extreme case of a covalent bond in which the Cl atom has hogged both electrons completely.

In practice, there is no such thing as a pure ionic bond, for no atom ever gains complete control over the electrons in the bond, and in fact AgCl is appreciably covalent. Therefore, whatever line we take on the degree of covalency, we can argue that the reaction is an example of Lewis acid-base complex formation. The same is true of all precipitation reactions. Therefore, one of our classes of reactions can be discarded (except for practical purposes): a precipitation reaction is merely one manifestation of Lewis acid-base behavior.

All reactions now appear to belong to one or another of three classes: acid-base, radical, and redox. We can tabulate the classes in a slightly different way that emphasizes the electronic events that accompany the reactions:

- Electron pair sharing: Lewis acid-base reactions
- Electron pair formation: radical reactions
- Electron transfer: redox reactions

We have arrived at where we asserted in Chapter 2 that chemical explanations should lie: in the behavior of electrons. All the phenomena of the world, insofar as they are transformations of one substance into another, are consequences of one of these three types of process.

However, there are still problems with accepting these three processes as truly fundamental. First, Lewis acid-base and radical reactions have a number of similarities. In some sense, radicals behave like acids and bases: they form the analog of complexes (when they combine), and they can displace one radical from another (when a radical attacks an ordinary molecule). Is there a way of expanding the definition of an acid and a base to capture radicals too? Another difficulty is the generally unsatisfactory state of the definition of a redox reaction in terms of an artificial concept, the exaggeration of electron distributions.

Is there a slightly different definition of an acid and a base that expands the domain of reactions captured by the concepts? The historical drift that has characterized the evolution of the concepts of acid and base has been (in a certain germ line, at least) the increased focus on the role of electrons. First Arrhenius, with no clear role for electrons, then Brønsted, with a shift of attention to proton donors and acceptors, and then Lewis and his identification of the importance of the electron pair. Yet radicals are characterized by reactions that involve not two but *one* electron. Can we go one stage further and focus completely on a *single electron?* Can acids and bases be understood in terms of individual electrons?

As a suggestion (and just for fun), consider the following definitions:

A base is an electron.
An acid is a hole.

By hole, we mean the absence of an electron from a distribution of electronic charge—a vacancy in an orbital. These definitions are closely related to the considerably more elaborate proposals made by the Russian chemist M. Usanovich in 1939 (and largely ignored ever since). We shall call these species U-bases and U-acids.

We have two questions to face before we can be satisfied that we have improved the earlier definitions. Do these definitions include Lewis acids and bases (and hence, by implication, Brønsted and Arrhenius acids and bases)? Do they extend the range of the meaning of acid and base in a reasonable manner? At this point we are not so much concerned with the usefulness of the definitions—after all, the Arrhenius definition of an acid is still widely used and extremely useful. We are looking for the intellectual bedrock of chemical reactions, the minimal type of change that can be identified as a reaction, the simplest process that would illuminate the candle in its entirety.

Note, first, that the definitions transfer the burden of being a base to the electron itself, not to anything that donates it, as in the Lewis definition.

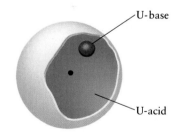

A fundamental acid (a U-acid) is a vacancy in an orbital (here, a single vacancy in an orbital that contains one electron). A fundamental base (a U-base) is a single electron.

That is analogous to the shift in focus from the Brønsted definition (in which the acid is the proton donor) to the Lewis definition (in which the proton itself is the acid). The Lewis definition of an acid more or less survives, because we must have a hole in something, so the electron pair acceptor is refined slightly to become an electron-accommodating hole.

It is trivial to confirm that all Lewis acids and bases are acids and bases under the new definition. In fact, each electron pair donor is the carrier of *two* U-bases, because each electron in the pair is capable of being involved in a reaction. Likewise, any Lewis acid is a double U-acid, because it has a vacancy for two electrons—a double hole in its electron distribution. A proton is in fact pure hole—there are no electrons here, but it is able to accommodate two. The whole of conventional acid-base theory is accommodated in the definitions.

The fundamental definitions have a greater domain of capture than the Lewis definitions because we do not need the presence of an electron pair or a completely empty orbital: a species that can donate a *single* electron in a reaction is the source of a fundamental base. That means that a radical is an U-base. However, a radical is also a U-acid, because it can accommodate an incoming electron (a U-base). Since a radical is both an acid and a base, it is appropriate to call it an amphoteric substance (from the Greek word for "both"). This term is already widely used in chemistry to denote substances that react with both acids and bases, so we must be careful to qualify it and call radicals U-amphoteric.

Now all the similarities between Lewis acid-base and radical reactions fall into place, for they are all reactions between U-acids and U-bases. At this stage, we appear to have collapsed our four classes of reaction down into two grand classes: acid-base reactions and redox reactions. All chemistry is captured in a nutshell of two classes of reaction!

But, of course, the question will naturally arise as to whether there is actually a *single* class of reaction. Have we yet unfurled the full strength of the definitions? (Of course, full strength could well be too great a strength, just as the classification "living things" is only of the broadest usefulness in biology.) Is it just possible that an oxidizing agent is a U-acid and a reducing agent is a U-base? If that is so, then all reactions in chemistry are acid-base reactions: we have uncovered the ultimate chemical yin and yang.

A reducing agent is an electron donor and is thus the source of a single U-base. An oxidizing agent is an electron acceptor and thus provides a single U-acid. In the oxidation of magnesium by chlorine, the magnesium is a source of two electrons (two U-bases), and when sodium is oxidized, it is the source of one electron (one U-base). When chlorine is reduced, it is the recipient of one U-base and hence is acting as a U-acid.

In this scheme of reactions, it is not necessary to try to keep track of electron drifts and the artificiality that they imply. All reactions are acid-base reactions, of a very fundamental kind. Hence, all the phenomena of the world are merely the reaction of one kind of acid with one kind of base. *That* is what the peculiar play of carbon amounts to.

CONTRIBUTIONS TO CHAOS | 4

*Mingling plumes of colored liquids model
the chaotic patterns of atmospheric
turbulence at the extremity of a wing. The
dispersal and mixing of the liquids
epitomizes the motive power of chemical
reactions that we explore in this chapter.*

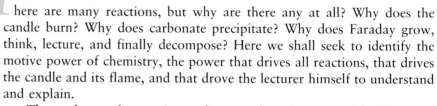

We come here to the philosophers, and I hope that you will always remember that whenever a result happens, especially if it be new, you should say, "What is the cause? Why does it occur?" and you will, in the course of time, find out the reason.

MICHAEL FARADAY, LECTURE 1

*T*here are many reactions, but why are there any at all? Why does the candle burn? Why does carbonate precipitate? Why does Faraday grow, think, lecture, and finally decompose? Here we shall seek to identify the motive power of chemistry, the power that drives all reactions, that drives the candle and its flame, and that drove the lecturer himself to understand and explain.

The tendency of a reaction to form products is governed by that most remarkable principle of nature, the "Second Law" of thermodynamics. The Second Law was established by Rudolph Clausius in 1850 and by William Thomson, Lord Kelvin, in 1851, but its applications to chemical reactions were not fully recognized until the work of Josiah Gibbs in the 1870s came to be appreciated at the turn of the century. Faraday was dead before he knew why his candle burned.

The Second Law is concerned with "spontaneous changes," the changes that have a natural tendency to occur without needing to be driven by an external agency. The term "tendency" has a great deal of sinister force in thermodynamics. It indicates that a system (a reaction mixture in our case) may be *poised* to undergo a particular natural change; however, there may be reasons why, though poised for change, the change is not realized. Hydrogen and oxygen have a natural tendency to form water (the reaction is spontaneous), but a mixture of the two can survive indefinitely until the floodgates of change are opened by a spark. Thermodynamics speaks only of the tendency to change; on the rate at which that change is achieved it is silent.

Certain changes are not spontaneous. Water has no tendency to decompose into hydrogen and oxygen, nor does carbon dioxide spontaneously crumble into oxygen and a layer of soot. However, in some cases nonspontaneous changes can be made to occur by *driving* a reaction in its unnatural direction. Faraday is famous for his success in one such unnatural pursuit,

Josiah Williard Gibbs (1839–1903).

Great disasters may stem from the tendency of the universe to become more chaotic. The ignition of hydrogen as the Hindenberg moored to its pylon in Lakehurst, New Jersey, in May, 1937 released the spontaneous tendency of hydrogen to combine with oxygen to form water. However, as we shall see, chaos can also be creative.

for one way of getting substances to change in an unnatural direction is by electrolysis, a technique that he studied in detail.

In this chapter we shall explore why a particular compound—such as carbon dioxide—has a tendency to be formed from the starting materials—the candlewax and the oxygen of the air. Similar arguments apply to the formation of water in a combustion, to the formation of carbonate ions when carbon dioxide dissolves in water, and to the precipitation of carbonates. All the chemical changes that Faraday conjured are manifestations of the spontaneity of reactions as expressed by the Second Law.

PRIMITIVE SPONTANEITIES

Thermodynamics is concerned with the transformations of energy, and the Second Law is, in brief, a summary of the tendency of matter and energy to disperse in a more disorderly form. That is, the law recognizes that nature has a natural tendency to lose whatever orderliness it currently possesses and to collapse into greater disorder: the natural tendency of change is toward decay. There are three primitive processes that illustrate what this

The molecules of a gas move chaotically, and spread away from their source. There is virtually zero probability that they will ever again all be found simultaneously in their original location. The primitive spontaneity this process illustrates is the chaotic dispersal of matter.

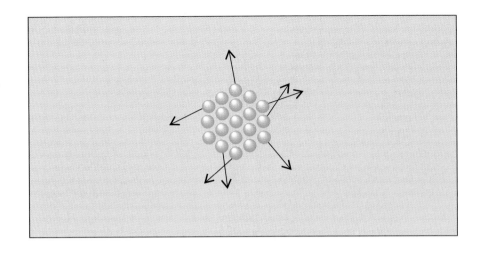

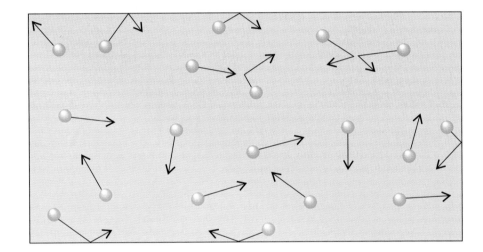

means. Exactly the same three primitive processes underlie all chemical change, so although they may seem trivial illustrations of the workings of nature, they are all that are required, with some circumspection of thought, to account for all the reactions of chemistry, and the chemical history of a candle in particular.

The first primitive spontaneity is the expansion of a gas to fill the available volume. We can think of carbon dioxide being created in the vicinity of the flame, and of the chaotic motion of the molecules resulting in their spreading through the room until the gas is dispersed uniformly. The molecules continue to move chaotically after their distribution has become uniform, but there is essentially no likelihood that they will ever again all simultaneously collect into the same initial region near the flame. That is, the spontaneous (and realized) direction of change of the system is the dispersal of matter. We can *make* the gas accumulate in any desired region, but, to do so, we have to do work on the system, such as by extracting the carbon dioxide from the air and releasing it close to the candle.

The second fundamental type of spontaneous change is one represented by the cooling of an object—a block of copper—heated in a flame. In this case, there is no mechanical dispersal of atoms, but there is a dispersal of the energy of the atoms that compose the block. The incessant jostling of the atoms of the block jostles the atoms in their immediate surroundings, they in their turn jostle their neighbors, or fly off far afield. Thus the energy of the heated block spreads chaotically as heat into the surroundings. There is very little likelihood that the jostling of atoms in the surroundings will lead to a significant accumulation of energy in the now cooled block, and that we would see its temperature suddenly rise and it begin to glow red hot again.

The third primitive spontaneity is slightly more elusive. The process can be illustrated by a ball that is heated by the candle. Energy flows into the ball (through the agency of jostling atoms), and if the candle flame is hot

Here, the direction of spontaneous change is clear at a glance: the particles of ink disperse through the water and become more disordered. However, in many cases the increase in disorder is harder to identify.

A red-hot block of copper cools to the temperatures of its surroundings, and the cooling is spontaneous. The reverse process, in which the copper becomes much hotter than its surroundings, has never been observed.

In the process of cooling, the chaotic "thermal motion" of the atoms in the hot body causes them to jostle their neighbors in the surroundings, and drive those atoms into motion. In this way, the energy spreads irreversibly away from the object. The primitive spontaneity this process illustrates is the chaotic dispersal of energy.

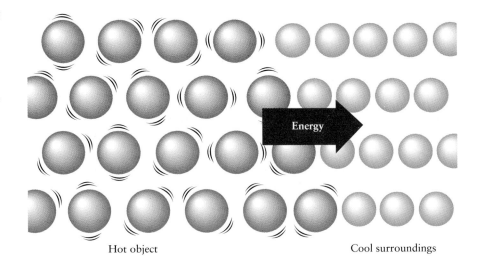

Hot object   Cool surroundings

enough, enough energy might flow in for the ball to be able to leap high into the air. But universal experience with balls heated by candles is that they do not suddenly fly upward: they just get hotter. The underlying reason is that although enough energy may flow into the ball, the motion of the ball's atoms is uncorrelated—it is higgledy-piggledy. For the ball to fly skyward, the atoms must be moving in a *correlated* manner—all upward at the same time. Similarly, when a bouncing ball comes to rest, it does so because the organized motion of its atoms is randomized a little more on each bounce, and the net motion of the ball is eventually stilled. We summarize this by saying that there is a tendency for correlated motion to decay into uncorrelated motion.

For motion to be imparted to a ball, it is not sufficient that energy is transferred: the motion of the atoms of the ball must be made to occur in a correlated way, with them all moving simultaneously in the same direction. When a ball bounces and comes to rest, this correlated motion is lost irreversibly. The primitive spontaneity this example exposes is the chaotic dispersal of the correlation of motion.

         CHAPTER FOUR

At this stage we can see that there are three reasons why a change may be spontaneous:

- The change may correspond to the dispersal of matter.
- The change may correspond to the dispersal of energy.
- The change may correspond to the loss of correlation of motion.

We shall see that these three primitive tendencies toward disorder account for all the chemical processes in the world. The tendency toward disorder is the motive power of change in all its forms, and in particular it is the true fuel of the candle (and of that more elaborate candle, Faraday himself).

ENTROPY AND THE SECOND LAW

The notion of disorder in the universe and the spontaneous tendency to ever greater chaos is put on a firm and quantitative foundation by the concept of "entropy." As far as we need be concerned, entropy is synonymous with disorder: as disorder increases so too does entropy. Should disorder decrease, then so too would entropy. Like all good concepts of the physical sciences, the concept of entropy can be made precise, unambiguous, and quantitative, but we do not need such a cathedral where a hut will do.

With the concept of entropy in mind, we can understand the force of Clausius's remark that

> *The entropy of the universe tends to increase.*

This crisp aphorism identifies the mainspring of change with the potency of chaos and is, for our purposes, a succinct statement of the Second Law. It is a more erudite, less picturesque (but potentially quantitatively powerful) way of saying that the universe tends to decay into disorder and chaos. It follows from our discussion of the three primitive spontaneities that there are three contributions to the entropy:

- Entropy increases as matter disperses.
- Entropy increases as energy disperses.
- Entropy increases as the correlation of motion is lost.

In our examples of primitive spontaneities, we have seen spontaneous *physical* changes—changes that do not change the substance (like dispersing and cooling). Now we must make the transition to *chemical* changes—changes that do change the substances—and observe that all chemical changes likewise occur in the direction of increasing entropy. That is, we must come to understand that whenever a chemical reaction occurs, there

As the disorder of the universe—a grand name for the system of interest and the surroundings with which the system can exchange energy—increases, the entropy increases.

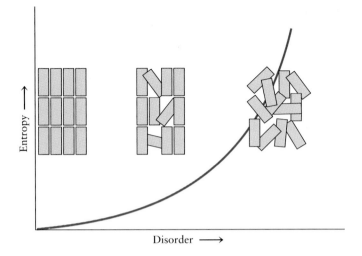

Entropy →

Disorder ⟶

has been an increase in the disorder of the universe—matter or energy may have become dispersed, or the correlation of motion may have become lost. That increased disorder may be deeply buried and quite difficult to discover, because we have to judge whether the new arrangement of atoms is, in a certain quantitatively precise sense, more disorderly than in the reactants, and we have to take into account the energy that may have accumulated in or been lost from the reaction as it took place. Indeed, the increase in entropy is not always to be found in the reaction itself. Sometimes we have to look far away from the site of the reaction, sometimes a hundred million miles away, for one process (nuclear fusion on the sun) may drive a reaction (the photosynthesis that carbon dioxide might undergo) here on earth.

THE FIRST CHEMICAL SPONTANEITY: DISSOLVING

We shall now explore some chemical reactions and become familiar with the kind of thinking needed to discover the underlying trend to chaos. The first reaction is the dissolution of a solid (remember that in Chapter 3 we saw that dissolution is a special case of a Lewis acid-base reaction).

Suppose we considered the dissolution of sodium carbonate, Na_2CO_3, in water. If we were to consider the motive power of change as being the tendency of matter to disperse, then we would apparently have little difficulty in understanding the solid's solubility. We would conclude that the compound dissolves because in the solid state the ions are closely packed together, but in solution they are dispersed in a disorderly manner. Hence, it

CHAPTER FOUR

would seem that there is a tendency for the solid to dissolve in water, for dispersal of matter is a process that has a natural tendency to occur.

However, we must not rush into thinking that we now understand why a substance dissolves. According to this description, *everything* should be soluble in anything else, which is not the case. Indeed, according to this argument, we do not even need a solvent—if a solid simply vaporized in a container, there would be an increase in entropy! Why do not all substances simply blow away as individual atoms or ions?

When considering changes in entropy, we must always consider the entire universe. Therefore, although there may be an increase in entropy in the part of the world we have focused on, there may be a decrease elsewhere, and the net change might not be spontaneous. Conversely, even if there is a local decrease in disorder, there may be a sufficiently great increase in disorder elsewhere that is coupled to the change we are observing, with the result that *overall* there is an increase in disorder when the combined process occurs. That is, we must always be very circumspect when dealing with entropy, and make sure that we have taken into account all the contributions that may be coupled to the process under consideration. It is for this reason that the descent into universal chaos may effloresce locally as a constructive process, as when the enzymes in our bodies synthesize a compound, or a chemist in a laboratory builds a structurally complex molecule from simpler parts. The mainspring of the world may be descent into chaos, but the gearing that links that spring to its outward manifestations is so intricate that locally the descent may appear to be ascent.

To understand why solids do not vaporize spontaneously into thin air, we must ask what changes in entropy we have ignored. We have concentrated on the dispersal of matter but have ignored the dispersal of energy. For solid sodium carbonate to blow away as a dust of ions would require an enormous influx of energy to rip the ions apart against the grip of opposite charges. That influx is a *localization* of energy—the equivalent of a cool copper block spontaneously glowing red hot—which leaves the disorder of the universe much reduced. Thus, sodium carbonate does not blow away as a dust of ions, for that would leave the universe *more* orderly than before.

Now suppose that there is water present. A feature of a water molecule that we now need to bring out is its "polarity," the fact that its atoms carry electric charges. The charge distribution arises from the electronegativity of the oxygen atom, which attracts the electrons in the O—H bonds and thereby creates a molecule that has a region of slight negative charge on the oxygen atom and regions of slight positive charge on the two hydrogen atoms. The slight difference in charge is a ripple on the surface of the overall distribution of electrons, but it is a ripple with a tidal wave of consequences, for it accounts for the composition of the oceans, the emergence of landscapes, and the workings of biology.

An H_2O molecule is polar in the sense that its centers of positive and negative charge are not coincident; the polarity arises from the difference in electronegativities of the hydrogen and oxygen atoms. The region rich in negative charge (the oxygen atom) can mimic an anion, and the regions thin in electron density and hence rich in positive charge (the hydrogen atoms) can mimic a cation.

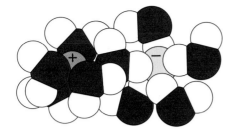

Water molecules can aggregate around ions by the process called "hydration." This hydration reduces the difference in energy between the ions in the crystal and the ions in the solid, so that less energy needs to be imported from the surroundings when the ions go into solution (in fact, in many cases energy can actually be exported to the surroundings).

The slightly negatively charged oxygen atoms of an H_2O molecule can mimic the anions in the original sodium carbonate crystal, and interact favorably with the Na^+ cations. Likewise, the slightly positively charged hydrogen atoms in an H_2O molecule can mimic the cations, and interact favorably with the CO_3^{2-} anions. Because the interaction between the water molecules and the ions is so strong, very little energy need flow into the vessel when the solid dissolves and the ions separate. Consequently, the presence of the solvent eliminates a potentially considerable decrease in entropy that would accompany a massive localization of energy. Now the principal change in disorder is the dispersal of matter, and the net effect when that occurs is an increase in entropy. Hence, dissolution in the strongly polar solvent water is a spontaneous process.

Water and other solvents soothe the process of dissolution by reducing the need to localize energy. However, water is not always successful as a solvent, for it does not always sooth enough. In some case (for example, in calcium carbonate and even more in granite), the interactions between ions in the solid are so strong that the interactions with the water molecules do not mimic them in strength particularly well, and a large influx of energy would be needed. That localization of energy is so improbable that we report that the solid is insoluble in water and observe that our landscapes are craggy: paradoxically, mountain ranges and buildings survive on account of the tendency of the world to become disordered.

THE SECOND CHEMICAL SPONTANEITY: COMBUSTION

Now we turn to Faraday's real but unrecognized problem: Why did his candle burn? The ostensible fuel of his candle was wax, which we have remarked consists of hydrocarbon molecules with long carbon chains. It will be easier to focus our initial remarks on the combustion of a much simpler molecule, that of methane, CH_4, a major component of natural gas. (Indeed, had Faraday presented his lectures now, he might well have dealt instead with the chemical history of natural gas, and have replaced his discourse on a candlewick with one on the thermoluminescent properties of a gas mantle.)

The overall combustion reaction that occurs when we burn natural gas in a plentiful supply of air is

$$CH_4(g) + 2O_2(g) \longrightarrow CO_2(g) + 2H_2O(g)$$

The question to consider concerns a dissymmetry of nature: Why does this reaction generate products (and heat) abundantly, but a mixture of carbon dioxide and water vapor has no tendency to form methane in the reverse reaction

$$CO_2(g) + 2H_2O(g) \longrightarrow CH_4(g) + 2O_2(g)$$

That is, why is the combustion reaction spontaneous, but not its reverse?

There is not a great deal of difference in the complexity of the reactants and the products of the methane combustion reaction, since three small gaseous molecules are replaced by three other small gaseous molecules. Therefore, if we improperly confined our attention to the disorder of the matter, we would conclude that in the combustion reaction there was little change in the disorder of the universe as it proceeded, and hence little reason why the combustion should occur at all. (Detailed calculation shows that there is in fact a small *decrease* in disorder.) However, our experience with

The combustion of natural gas, either on a small scale as in a domestic furnace or on a much larger scale as from this ocean oil-production platform, is a combustion reaction that is driven by the disorder that results from the spread of particles and, principally, the dispersal of energy.

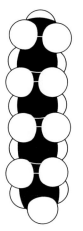

Octane, C_8H_{18}

dissolving has taught us that we should be circumspect and take into account the chaos that may stem from the dispersal of energy. When a combustion reaction occurs, it releases a lot of energy as heat into the surroundings as strong C=O and O—H bonds are formed to replace the weaker C—H bonds. That release amounts to a very considerable dispersal of energy, and the disorder it generates swamps the small change in disorder of the newly formed products. Hence, overall the universe is more chaotic after methane has burned than it was before, and so the combustion reaction is spontaneous.

We can now also begin to see one of the reasons why gasoline is such an effective fuel. A compound representative of the many present in the complex mixture that we purchase at pumps is the hydrocarbon octane, C_8H_{18}. The combustion reaction of octane vapor is

$$2C_8H_{18}(g) + 25O_2(g) \longrightarrow 16CO_2(g) + 18H_2O(g)$$

and for every 27 molecules of reactants that react, 34 molecules of products are formed. Unlike for methane, where the combustion led to hardly any change in the structural order of matter, in the combustion of octane there is a disintegration of structure: the fuel molecule is a cluster of 26 atoms bound together as a compact unit, and the products consist of many independent triatomic molecules. So, the spontaneity of the reaction is due both to the considerable increase in entropy as energy disperses and also to the increase in entropy that stems from the chaotic collapse of the matter itself. Hence, octane and its analogs pack a considerable punch, which drives the reaction forward and—coupled to the earth through pistons, gears, and wheels—drives us forward too.

Faraday's candlewax was an effective fuel for the same reasons, for its lengthy hydrocarbon chains underwent extensive disorder-creating disintegration once he had kindled the flame. Moreover, the combustion gave out heat, and so it also stirred up thermal disorder in the surroundings. The candle burned, and the world sunk a little further into chaos.

THE THIRD CHEMICAL SPONTANEITY: ENDOTHERMIC PROCESSES

There are some reactions that appear to the untutored eye to run against nature, absorbing energy, not releasing it. The existence of these reactions was the source of much perplexity before chemists had acknowledged that it was entropy that they should scrutinize as the criterion of spontaneity, not energy. They saw energy flooding into their reaction vessels in these "endothermic" reactions, and were bewildered by their suspicion that the reactants were falling upward in energy as they formed products. How, they wondered, could a system spontaneously rise up to higher energy?

An example of an endothermic chemical reaction is the decomposition of calcium carbonate, the lime-making reaction that we saw as an example of a Lewis acid-base reaction in Chapter 3, which occurs at temperatures above about 800°C:

$$CaCO_3(s) \longrightarrow CaO(s) + CO_2(g)$$

Why does the reaction have a spontaneous tendency to roll uphill to a state of higher energy?

The spontaneity of an endothermic reaction is not perplexing once we acknowledge that the direction of spontaneous change is determined by entropy, not energy, and in particular by the *overall* disorder of the universe. Since energy becomes localized in the system when the reaction takes place, the entropy of the surroundings is decreased. However, in the reaction, a gaseous molecule is produced for each carbonate ion that is consumed, and because a chaotic gas is produced from a solid, the *matter* is very much more chaotic at the end of the reaction than it was at the beginning. Indeed, the increase in disorder of the matter may be so great that it overcomes the decrease in disorder of the surroundings, and there is an overall increase in disorder.

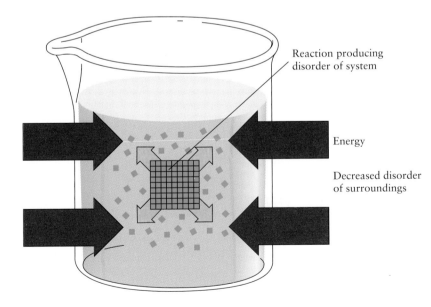

Reaction producing
disorder of system

Energy

Decreased disorder
of surroundings

An endothermic reaction is spontaneous when the disorder generated in the system is greater than that eliminated from the surroundings as energy flows into the reaction. Overall, the reaction generates disorder and hence is spontaneous.

An endothermic reaction, such as this reaction between barium hydroxide octahydrate and ammonium thiocyanate, draws in energy from the surroundings. So much energy may be absorbed that any water that happens to be in contact with the reaction is frozen, as we see here from the frost forming on the surface of the beaker.

Reactions that are driven by the dispersal of matter give rise to the concept of an "antifuel." An antifuel—not that anyone actually calls it that—is any substance that reacts endothermically, and hence absorbs energy from the surroundings, so cooling them. Although it is possible to envisage purely chemical refrigerators, the counterparts of furnaces, powered by antifuels, they would be uneconomical in practice except for very special applications. There are only one or two practical uses of chemical refrigerators, perhaps the best known being the "instant cooling packs" that are applied to muscle sprains. The antifuel of these packs is ammonium nitrate, NH_4NO_3, and water: the solid is held in a tube that is immersed in the water of the pack. When the pack is needed, the tube is broken and the solid dissolves. The dissolution of ammonium nitrate is a significantly endothermic process, largely because the strongly attracting NH_4^+ and NO_3^- ions must be separated; however, the dissolution causes considerable disorder as the ions break down the orderly arrangement of the water molecules in pure water, and this increase in the disorder of matter is greater than the decrease in disorder as energy floods into the reaction and cools the surroundings.

We can now go on to see why endothermic reactions may be spontaneous at high temperatures even if they are not spontaneous at low temperatures. Consider, once again, the thermal decomposition of carbonate ions in the lime-making reaction. The thermodynamic effect of elevated temperature is akin to the soothing effect of a solvent in a dissolution: it reduces the reduction in disorder in the surroundings caused by the influx of energy, and so lets the matter find its natural destiny—the chaos typical of the products (such as carbon dioxide and lime).

All spontaneous endothermic processes derive their spontaneity from the increase in disorder of the matter in the system, which overcomes the decrease in disorder of energy in the surroundings. All reactions, whether energy releasing (exothermic) or endothermic, are driven by the same tendency for the universe to become more disordered. The tendency to chaos is truly the mainspring of chemistry. Faraday's flame, and all his biochemical actions as a human being, were driven by the tendency of matter and energy to disperse.

COUPLED SPONTANEITY

A central aspect of chemistry is that even a nonspontaneous chemical reaction can be made to occur. Faraday used the technique unwittingly in his lectures when he prepared hydrogen and oxygen by passing an electric current through a solution, the technique known as electrolysis. He knew that hydrogen and oxygen combine to form water; he knew intuitively and we know thermodynamically that the reaction

$$2H_2(g) + O_2(g) \longrightarrow 2H_2O(l)$$

is spontaneous (and is explosively fast once some radicals—from a spark—are present). Faraday, the master of electrolysis, also knew that he could undo the reaction electrically, and convert water back into hydrogen and oxygen

$$2H_2O(l) \longrightarrow 2H_2(g) + O_2(g)$$

by passing an electric current through water. However, because the necessary conceptual framework was not available to him, he must have failed to appreciate that with electrolysis he was driving a reaction in its nonspontaneous direction. The forward reaction (the formation of water) is spontaneous since it increases the disorder of the universe by allowing energy to spread into the surroundings. Faraday's achievement of the reverse reaction, the decomposition of water, appears to be flying in the face of nature, since it must *decrease* the disorder of the universe by the same amount as the forward reaction increases the disorder! Did Faraday know something that we do not?

No, Faraday knew even less than we do about spontaneity and entropy: nature had taken the reaction out of his hands, and by a typical sleight of

The apparatus Faraday used in his lecture to demonstrate the electrolysis of water. When a current is passed, bubbles of hydrogen gas appear at the cathode and bubbles of oxygen gas appear at the anode. The volume of hydrogen produced is twice that of oxygen.

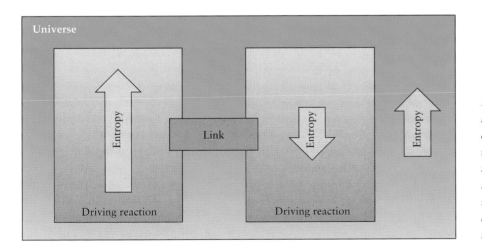

If a reaction that is highly spontaneous (in the sense of resulting in a large increase in total entropy), as represented by the exothermic reaction on the left, is coupled to a reaction that is nonspontaneous (for instance, because it is quite strongly endothermic), the former reaction may drive the latter in its unnatural direction because overall there is an increase in total entropy.

hand had achieved an overall increase in order despite causing the influx of energy into the sample as the H—O bonds in water were severed. The crucial point is that *a strongly spontaneous reaction can drive another reaction in the latter's nonspontaneous direction*. The mechanical analogy is that of a falling weight which, through a system of pulleys, can raise a lighter weight. If we could see only the lighter weight winging its way upward, we might feel justified in celebrating a miracle (or at least the demise of the Second Law). However, as soon as we see that the light weight is coupled to a heavier falling weight, the mystery disappears and the miracle is confounded (and the Second Law preserved). *All* apparent disorder-decreasing reactions are in fact reactions that are driven by the generation of more disorder elsewhere. So long as the two reactions are coupled in some way, a reaction that produces a lot of disorder may be able to drive another reaction in its unnatural direction, the net result being the generation of disorder in the universe even though *locally* (in the electrolysis vessel and in its immediate surroundings, for instance) disorder has declined.

We now know that when, in his lecture, Faraday passed an electric current into his electrolysis cell, he was pumping in electrons through one electrode (the cathode) and sucking them out through the other (the anode). The electrons he supplied attached to hydrogen ions that were close to the cathode, and reduced them to molecular hydrogen:

$$2H^+(aq) + 2e^- \longrightarrow H_2(g)$$

In other words, a cathode is a very powerful reducing agent. (The electrolysis goes faster if the water is acidified, for that increases the otherwise very low concentration of hydrogen ions that are present in pure water.) The electrons he dragged out of the water through the other electrode were stripped off the water molecules close to the anode, which acts as a powerful oxidizing agent:

$$2H_2O(l) \longrightarrow O_2(g) + 4H^+(aq) + 4e^-$$

That is, as electrons were sucked out of the liquid, the water was oxidized to molecular oxygen. As the net result of pushing electrons into the cell through the cathode and dragging them out through the anode, Faraday achieved the reduction of water to hydrogen at the cathode and its oxidation to oxygen at the anode.

In his lecture, Faraday used a "voltaic pile" as the source of the electric current. The voltaic pile was invented by the Italian Alessandro Volta in 1799 (and the device helped, incidentally, to scotch the view that animal tissue was essential to the generation of electricity), and in the form used by

The voltaic pile used by Faraday in his lectures. It consists of a column of alternating copper and zinc disks separated by paper soaked in acidified water. It generates a current of electrons that emerge from the zinc electrodes, pass through an external circuit, and then reenter the pile at the copper electrodes.

CHAPTER FOUR

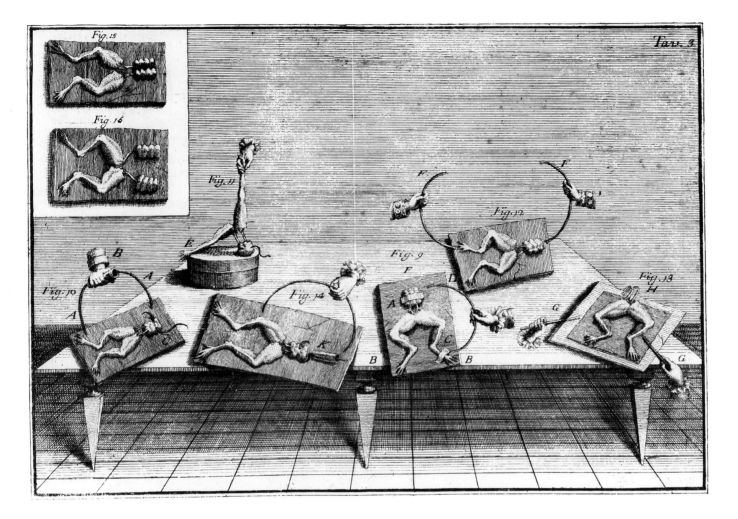

Faraday consisted of a series of alternating copper and zinc disks that are separated from each other by cloth that had been dampened with acidified water. (Count Volta originally used a more aristocratic pile of silver and zinc.)

Both the voltaic pile and the cell in which the electrolysis is taking place have essentially the same construction (consisting of electrodes separated by a conducting medium), and both are "electrochemical cells." An electrolysis cell is an electrochemical cell in which a reaction is brought about by an electric current, and its counterpart, a "galvanic cell," is an electrochemical cell in which a reaction generates an electric current. The latter is named after the Italian physiologist Luigi Galvani, who in 1786 had identified a

Galvani's initial observation was that dead frogs fixed by brass skewers to an iron fence underwent convulsions and later made a more systematic study of the effect. He believed that a new kind of electricity was generated by the animal, but Volta's work disproved this assumption.

The overall reaction given in the text occurs when a piece of zinc is placed in a solution of copper sulfate (which contains Cu^{2+} ions). A redox reaction occurs in which zinc is oxidized to Zn^{2+} ions and Cu^{2+} ions are reduced to copper metal, which plates out on to the surface of the zinc.

connection between muscular convulsions and electricity, and whose work had stimulated Volta to begin his own.

The ability of the voltaic pile to generate an electric current depends on the properties of electrons and the workings of the Second Law. The driving power of the reaction in the voltaic pile is the spontaneous tendency of zinc metal to reduce copper ions to copper metal. That is, there is a net increase in disorder in the universe when zinc atoms in the zinc electrode release electrons (and hence are oxidized):

$$Zn(s) \longrightarrow Zn^{2+}(aq) + 2e^{-}$$

and the copper ions accept them (and hence are reduced):

$$Cu^{2+}(aq) + 2e^{-} \longrightarrow Cu(s)$$

The net reaction is therefore the redox reaction

$$Zn(s) + Cu^{2+}(aq) \longrightarrow Zn^{2+}(aq) + Cu(s)$$

The crucial feature of the overall reaction is that electrons are released by the zinc and are taken up by the copper ions.

The clever trick adopted in a voltaic pile (and all galvanic cells) to achieve the production of electricity from a chemical reaction is to separate in space the sites of oxidation and reduction. Oxidation takes place at each of the zinc disks, where zinc atoms lose electrons, and the ions so formed break away from the metal and go into the solution that soaks the paper. The electrons travel through a wire connecting the zinc disks to the copper disks, and attach to the copper ions that are in solution close to the latter disks. It is this flow of electrons that constitutes the "electricity" produced by the cell and which was used by Faraday for his electrolysis demonstration. All electrochemical cells act in much the same way, from the large batteries used in automobiles (where the reactions are the oxidation and reduction of lead and its oxides) to the lithium batteries used in heart pacemakers.

Now we can see how one reaction, the reduction of copper ions by zinc metal, can drive another reaction, the decomposition of water, in its nonspontaneous direction. We simply link the two reactions by including the electrolysis cell in the external circuit of the galvanic cell. Although the decomposition of water by electrolysis causes a decrease in the disorder of the world, the reduction of copper ions in the voltaic pile by zinc metal occurs with such a large increase in disorder that the net effect on the world

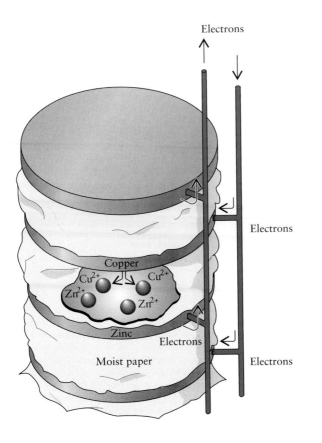

Electrons

Electrons

Copper

Cu^{2+} Cu^{2+}

Zn^{2+} Zn^{2+}

Zinc

Electrons

Moist paper Electrons

The electric current generated by the voltaic pile arises from the loss of electrons by the zinc atoms and their gain by the copper ions in the damp paper separating the disks. That is, the zinc undergoes oxidation in one region of the pile and the copper undergoes reduction in another region. The overall reaction is the same as in the preceding illustration, but the transfer of electrons between the centers of oxidation and reduction constitutes the flow of electric current.

is an increase in its disorder, and the joint, coupled pair of reactions is spontaneous overall. If, as we now might do, we replaced the voltaic pile by a distant power station, exactly the same principle would apply, but the generation of disorder we use for our local delight would be the combustion of oil or coal, the fissioning of nuclei, or the thunder of a crashing waterfall.

The same principles apply to all reactions that occur because they are driven by other, more spontaneous reactions. The mechanism of human life is a consequence of exactly the same principle as is demonstrated so elegantly by the voltaic pile and the electrolysis cell. The achievement of growth, for instance, is a local reduction in disorder, and it is driven by reactions that generate even more disorder elsewhere in the body. Thus, one aspect of growth is the synthesis of proteins (from amino acids). The formation of protein molecules is accompanied by a considerable reduction in entropy as all their small amino acid molecules are linked together in a genetically controlled sequence. However, each step is linked to a reaction in which one or more biological molecules (ATP) partly decomposes (into

A protein molecule (this one is ribonuclease) consists of a long chain of amino acids bonded chemically together. An amino acid is a molecule of the form NH₂CHRCOOH, where R is a group of atoms, and a protein is a "polypeptide" that consists of chains of the form —NHCHRCO—NHCHR'CO—NHCHR"CO— and so on, with the identities of the R groups laid down by the genetic code.

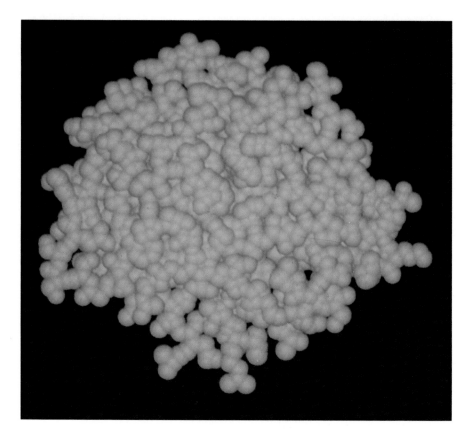

ADP). The increase in universal disorder brought about by that decomposition is so great that it more than compensates for the decrease in entropy that accompanies the formation of a polypeptide.

The partial decomposition of the ATP to ADP is linked to the formation of bonds between the amino acids by a system of enzymes, and *overall* the construction of proteins is spontaneous because the coupled reactions collectively lead to a collapse of order. The gearing of the reactions is such that the polypeptide is built at the expense of greater disorder elsewhere. To sustain these reactions, we need to eat, for we must rebuild the ATP from the ADP, using the energy of reactions stemming from respiration and digestion. In a green leaf—one of the sources of the food that fuels us—the rebuilding of ATP from ADP is achieved by photosynthesis, and there the distant waterfall that drives the reaction is nuclear fusion in the sun. All life is the outgrowth of this intricately geared collapse into cosmic chaos.

CHAPTER FOUR

REACTIONS AT EQUILIBRIUM

Now let us shift our attention to another intriguing aspect of a chemical reaction. It is often the case that a reaction seems to stop even though some reactants remain. For example, after carbon dioxide dissolves in water and forms carbonic acid, the acid surrenders a proton to a water molecule in the reaction that we saw in some detail in Chapter 3:

$$H_2CO_3(aq) + H_2O(l) \longrightarrow HCO_3{}^-(aq) + H_3O^+(aq)$$

However, by measuring the concentration of H_3O^+ ions in the solution it is known that not all the H_2CO_3 molecules surrender a proton. The reaction—the proton transfer—proceeds so far, and then stops; it is as though the formation of product strangles the reaction. There is a further peculiarity: analysis of the composition at this apparent dead end of the reaction shows that, whatever the starting composition, at 25°C the proton-transfer reaction comes to a halt when the concentrations of products and reactants reach particular values. Whatever the initial composition and temperature, in due course the reaction seems to come to an end—the reaction has reached "equilibrium"—when the concentrations have a specific relationship to each other. (Specifically, the concentrations [X] satisfy the relation $[HCO_3{}^-][H_3O^+] = K \times [H_2CO_3]$, where K is the "equilibrium constant" of the reaction at the prevailing temperature.) Why does a reaction apparently halt at a characteristic set of concentrations? Why does a reaction apparently lose its spontaneous character?

Like so much in chemistry (that subtle science), appearances can be deceptive. The reaction has not in fact stopped—it is still going on; it simply

The reaction that is used to power many changes in a biological cell—including the synthesis of proteins and the propagation of nerve impulses—is the reaction in which ATP (shown here on the left) loses a phosphate group and becomes ADP (on the right). The ATP is then rebuilt using energy from sources such as respiration and photosynthesis.

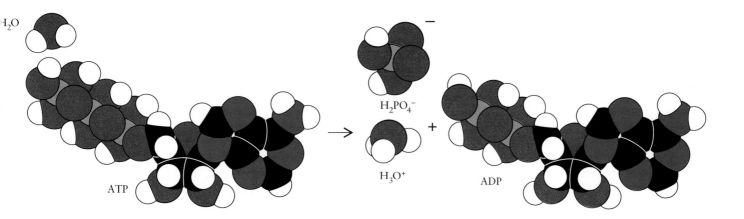

seems to have come to a stop. Reactions never actually stop. To understand this important point, we must appreciate that most processes in the world—chemical reactions included—can run in either direction. Therefore, not only is it possible for H_2CO_3 and H_2O to react and form HCO_3^- in the "forward" reaction shown above, but the HCO_3^- ions can reform the starting materials in the reverse of that reaction.

When we refer to a reaction in progress, we mean one in which the forward reaction is occurring more rapidly than the reverse reaction, so that there is a net formation of product, of HCO_3^- ions in this case. As the composition approaches equilibrium, the forward reaction slows (because there are fewer reactants remaining) and the reverse reaction goes more quickly (because there is more material to undergo decomposition), and the net formation of products proceeds more slowly. At equilibrium, the rate of the forward reaction is exactly matched by the rate of the reverse reaction, and there is no net formation of products or decomposition into reactants.

The state of equilibrium that we have described is a living equilibrium, in which the forward and reverse reactions are proceeding but are balanced. It is called a "dynamic equilibrium" to distinguish it from the static equilibrium of a balanced see-saw. The state of dynamic equilibrium can be likened to a busy store, where the number of customers entering through its doors is equal to the number leaving. Static equilibrium is like a store that has gone bankrupt. The physical analogy of a dynamic equilibrium is the hot copper block that has cooled spontaneously to the temperature of its surroundings (and hence has attained "thermal equilibrium" with them): the transfer of energy between the block and its surroundings continues, but the outflow of energy from the block is now balanced by the inflow of energy from the surroundings. The fact that chemical equilibrium is dynamic is of immense importance, because it means that an equilibrium mixture is a responsive, living thing, not just a dead lump of mixed substances.

CHEMICAL SELF-STRANGULATION

To discover why a reaction undergoes self-strangulation, we should think of the disorder of the universe when in the solution there are only H_2CO_3 and H_2O molecules, and of the disorder of the universe at the end of a hypothetical reaction that has gone to completion, when all the H_2CO_3 molecules have become HCO_3^- ions. It is known that at 25°C the disorder of the universe is greater when all the carbonic acid is present as H_2CO_3 molecules than when it is present entirely as HCO_3^- and H_3O^+ ions. So it appears that there is no spontaneous tendency for proton transfer, let alone for the reaction to be self-strangling! Somewhere, though, there must be a virus hidden in our argument, because some proton transfer certainly occurs, and after a certain concentration of HCO_3^- ions has been reached, those ions eliminate the spontaneity of further reaction.

As always when dealing with entropy, we have to be alert. What we need in this case is a contribution to the entropy that is not present when we have only H_2CO_3 molecules or only HCO_3^- and H_3O^+ ions, and that reaches a maximum at an intermediate stage of the reaction. The overall entropy will then be determined by three contributions: the change in the species present, the dispersal of energy, and this new factor.

It is the product itself that is responsible for the additional contribution to the entropy. As HCO_3^- is produced from the reactants, the reaction system becomes a mixture, and therefore becomes more disorderly than when there are only H_2CO_3 molecules present. This disorder initially increases as proton transfer occurs and the solution consists of all three species H_2CO_3, HCO_3^-, and H_3O^+, but it falls again as the mixture becomes depleted in H_2CO_3 and the composition of the mixture simplifies. As shown in the illustration, this extra contribution to the entropy causes the disorder of the universe to rise from its initial (only H_2CO_3) value, pass thorough a maximum after a tiny proportion of protons have been transferred, and

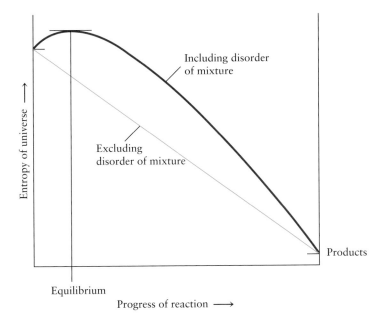

The contribution to the disorder of a system that arises from the presence of several species changes as the reaction proceeds. It is zero when only one species is present, rises to a maximum when several species are present in the middle of a reaction, and then decays again as the reactant ceases to make a contribution to the disorder when it has been almost entirely consumed. This contribution to the entropy adds to the other two (which arise from the change of species and the release of energy), and leads to a maximum in the graph of entropy against the progress of the reaction.

then decline to the value typical of complete proton transfer. The reaction has a spontaneous tendency to form reactants until the composition has reached the point corresponding to maximum universal disorder, when its forward progress is no longer spontaneous. The net reaction ceases and settles into dynamic equilibrium. The product has both enabled and strangled its own formation.

REACTIONS THAT DO AND REACTIONS THAT DON'T

All reactions have a specific equilibrium composition that corresponds to the point at which, as a result of their occurrence, the disorder of the universe is maximized. In some cases, the reaction reaches equilibrium at a composition that, to the undiscerning eye (and in many cases to the exquisitely discerning), appears to correspond to pure products. This is the case when a piece of zinc is placed in a solution of copper sulfate, and the redox reaction

$$Zn(s) + Cu^{2+}(aq) \longrightarrow Zn^{2+}(aq) + Cu(s)$$

takes place, with the formation of a reddish-brownish-black deposit of copper. At room temperature, equilibrium is reached at a concentration that corresponds to a ratio of Zn^{2+} to Cu^{2+} ions of 2×10^{37} to 1. That is, if we had a solution of copper sulfate that occupied a sphere the volume of the earth, and enough zinc metal were added to it, at equilibrium there would be only a few dozen Cu^{2+} ions in the whole of the solution. With one more Cu^{2+} ion than at equilibrium, the universe would be less disorderly, just as it would be if there were just one ion less.

There are also some reactions that attain dynamic equilibrium when the reaction has barely begun. For example, suppose we started with an earth-size flask of a solution of zinc sulfate, with some lumps of zinc present. If we dropped a couple of hundred Cu^{2+} ions into the solution, about half of them would be reduced by the zinc and an equal number of Zn^{2+} ions would enter solution. Then the reaction would stop, for further production of Zn^{2+} ions would decrease the disorder of the universe. We would, very legitimately, say that the reaction "does not go."

Many reactions that do not go, do not go on these grounds. That is, the reactants are so close to the equilibrium composition of the reaction that the formation of a merest wisp of product strangles the reaction. The maximum in entropy of the universe lies almost infinitesimally close to the composition corresponding to pure reactants.

In all his chemical demonstrations, including the combustion of the candle and the decomposition of water by electrolysis, Faraday was building understanding by unknowingly contributing to the disorder of the world. In some cases he was able to unleash the spontaneity of a reaction with only a little technological intrusion, as when he kindled a flame. However, some of the reactions would proceed in an unwanted direction unless they were harnessed to a more powerfully spontaneous process; such coupling of reactions requires technological refinement, and its achievement occurred at a much later stage in the progress of civilization. A reaction of this kind is the electrolysis of water, which Faraday was able to drive by harnessing the flow of electrons generated by another reaction. All reactions, however, are open to strangulation, for in all cases there is a contribution to the disorder of the world that stems from the simultaneous presence of the reactants and the products, and maximum disorder—and its accompaniment, the cessation of overt reaction—occurs when all are present. All reactions, then, stem from the collapse of the world into ever greater chaos. Even Faraday's conjuring of the orderliness of understanding in the minds of his lecture audience was, deep down, a contribution to chaos.

THE BLOCK ON THE PATH | 5

Although many reactions have a tendency
to occur, they may not be able to proceed
until the temperature is raised or some
other stimulus is applied. The match shown
here bursts into flame when it undergoes
reactions stimulated into action by the heat
generated by friction.

It is curious to see how different substances wait—how some will wait till the temperature is raised a little, and others till it is raised a good deal.

MICHAEL FARADAY, LECTURE 6

*F*araday's candle waited. Even though the combustion of hydrocarbons is spontaneous, Faraday needed to ignite his candle to bring it awake. Once the flame had been established, the heat it produced sustained the reaction and could also be used to initiate the combustion of other substances. The conversion of wax to carbon dioxide and water is spontaneous, but at room temperature it proceeds so slowly that in practice there is no change at all and candles can be stored indefinitely. Only at the elevated temperatures characteristic of the flame does the spontaneity of the reaction become manifest, and the wax burns into its chemical destiny.

It is quite easy to identify the reason for the slowness of the reaction at room temperature: the atoms of the molecules that make up the candle are gripped together by strong bonds, and even though in due course energy will be released as stronger bonds are formed, the liberation of the atoms from their original bonds requires a considerable initial investment of energy. The atoms of the candle are poised for change, but unable to achieve it, like a lake held back by a dam. We say that there is a "kinetic" barrier to their rearrangement (because the study of the rates of reaction in chemistry is called "chemical kinetics"): the atoms do not have enough energy to change their partners despite their natural tendency to do so. Not merely Faraday's candle but all the forests of the planet, and all its life, are poised on the brink of conflagration, yet so strongly are their atoms gripped that the world does not tumble headlong into chaos but unfolds (except here and there) with a serene richness. To achieve a complex world, reactions wait.

Reactions vary in the energy investment they require, and hence they show a wide range of rates. The precipitation of insoluble calcium carbonate is virtually instantaneous, and the ions of the precipitate clump together immediately calcium ions and carbonate ions are present in the same solution. The neutralization of carbonic acid by a base is also almost instantaneous because protons are highly mobile. However, as is familiar from cooking, some substances can be left mixed for hours before an appreciable

CHAPTER FIVE

Wax platelets are extruded from the glands in the abdomen of a bee.

amount of product has formed, and sometimes no product at all is formed until the mixture is heated.

An example of a reasonably slow reaction is the analog of the biochemical processes that nature used to prepare the wax in Faraday's candles. Modern candles are typically paraffin wax—a white solid obtained from petroleum, which consists of a mixture of hydrocarbon molecules containing between 18 and 32 carbon atoms. Faraday's candles, like some ceremonial candles still used today, were made of tallow, spermaceti (while whales were still plundered), or beeswax, for in his time petroleum products were still something of a rarity. A true animal wax is a complex mixture of organic molecules, including long-chain hydrocarbons, carboxylic acids (compounds that contain the group —COOH), and esters formed by the reaction of a long-chain carboxylic acid and a long-chain alcohol (an alcohol is a compound that contains an —OH group attached to a carbon atom). Typical chain lengths are of the order of two or three dozen carbon atoms. The formation of the long-chain esters in beeswax is carried out under enzymatic control in the abdomens of bees, but the welding of one molecule to another to form an ester can be modeled under much simpler conditions simply by warming a mixture of the organic acid and the alcohol.

Then an "esterification reaction" such as

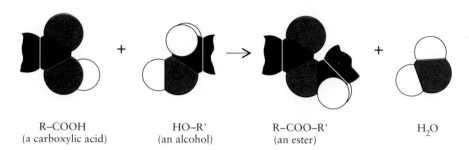

R–COOH	HO–R'	R–COO–R'	H$_2$O
(a carboxylic acid)	(an alcohol)	(an ester)	

takes place. Esterification reactions such as this (and also the true wax-producing reactions that they model) take place at a reasonably leisurely rate, and it may take many minutes and sometimes hours for the mixture to attain its ultimate equilibrium.

In this chapter, we explore some of the factors that determine the rates of the reactions that Faraday's flame unleashed upon the world: we see why some reactions are seemingly held back from immediate formation of product whereas others plunge headlong into their chemical destiny. The causes of the different rates of chemical reactions and our ability in many cases to exercise control over them—holding back the fast and spurring on the slow—will be recurring themes of the remainder of this account.

ARRHENIUS BEHAVIOR

It is almost always the case, and is true for the esterification reaction above, that the rate at which a product is formed increases with rising temperature. A very few reactions, bizarrely, go more slowly at higher temperatures, but that chemical byway is something we need not worry about until the next chapter. We take advantage of the sensitivity of reaction rate to temperature when we freeze food to slow the reactions leading to its decay or cook it to speed the release of components that add to its flavor and aroma—before it succumbs to the slower reactions that we interpret as putrefaction. The use of increased temperature to accelerate reactions is also one of the body's lines of defense against infection, for in a fever the body temperature rises in an attempt to kill the attacking bacteria by distorting the fine balance of rates of their biochemical reactions.

The variation of reaction rate with temperature conforms to a simple rule proposed by the Dutch chemist Jacobus van't Hoff in 1884 (in the first treatise on reaction rates, his *Études de dynamique chimique*) and interpreted by Svante Arrhenius in 1889. This rule states that the rate of many reactions can be expressed as an exponential function of the temperature (T):

$$\text{Rate} \propto e^{-T_a/T}$$

(T is measured on the Kelvin scale, with $T = 0$ at $-273°C$.) Reactions with rates that depend on temperature in this way are said to show "Arrhenius behavior." The quantity T_a varies from reaction to reaction and we shall call it the "activation temperature" of the reaction. The empirical significance of the activation temperature is that it indicates the sensitivity of the reaction rate to changes in the temperature. An activation temperature of zero signifies that the reaction rate does not depend on the temperature. An activation temperature of about 5000 K signifies that the reaction rate doubles when the temperature is raised from room temperature by about 10°C. A typical activation temperature is about 5000 K, but some reactions have an activation temperature close to zero, and some have activation temperatures of the order of 10^4 K or more. These values of the activation temperature are high compared with room temperature, which is usually less than 300 K, but are consistent with the slowness of many reactions under normal conditions (because $e^{-5000/300} = 6 \times 10^{-8}$). Raising the temperature to nearly 500 K, as happens in cooking in an oven, results in a thousandfold increase in reaction rate (because $e^{-5000/500} = 5 \times 10^{-5}$).

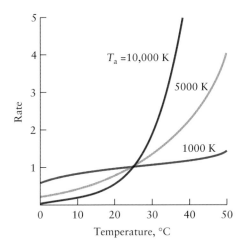

The higher the activation temperature of a reaction, the more sensitive is the rate to temperature.

COLLISIONS IN REACTIONS

The easiest way to understand the effects of increasing temperature is to consider the events taking place in different regions of a candle flame. The particular event that we need to imagine is a collision. For example, we saw in Chapter 3 that one of the processes in the flame is the collision of an OH radical with a fragment of a hydrocarbon molecule, and in the upheaval of the collision the radical may snip out a hydrogen atom and carry it off as an H_2O molecule. The collision of the two species is the key event that allows them to react, and it is the basis of a theory that explains the relation between rate and temperature.

The "collision theory" of reactions starts from a model of a gas in which all the molecules are in continuous, chaotic motion. The model is appropriate to the flame of our candle, for the flame is a turbulent gas of molecules, molecular fragments, and particles of incandescent soot. Collisions occur throughout the flame, and any given molecule typically collides

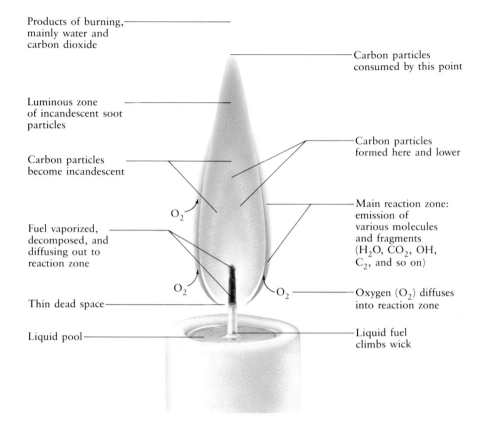

The structure of a candle flame and the principal processes occurring in each region.

Products of burning, mainly water and carbon dioxide

Carbon particles consumed by this point

Luminous zone of incandescent soot particles

Carbon particles formed here and lower

Carbon particles become incandescent

Main reaction zone: emission of various molecules and fragments (H_2O, CO_2, OH, C_2, and so on)

O_2

Fuel vaporized, decomposed, and diffusing out to reaction zone

O_2 O_2

Oxygen (O_2) diffuses into reaction zone

Thin dead space

Liquid pool

Liquid fuel climbs wick

with another several billion times a second. If every collision were successful, the reaction would be over in a few billionths of a second—as soon as the hydrocarbon fragments had left the wick—and would not take place throughout the volume the size of a typical flame.

As in most explanations of chemistry, we look to the energy requirements of the process for the explanation of puzzling behavior—in this case, for the reason why the reaction occurs so slowly compared to the rate at which collisions occur, and ultimately why the rate varies with temperature. Although there is a storm of collisions in the flame, most of the collisions are too feeble (particularly in its cooler regions) to result in reaction. The OH radical might collide with a hydrocarbon fragment, and perhaps start to draw out a hydrogen atom from it, but the collision might not provide enough energy for the extraction to go to completion. Instead, the OH radical bounces off the hydrocarbon, and the molecule remains intact.

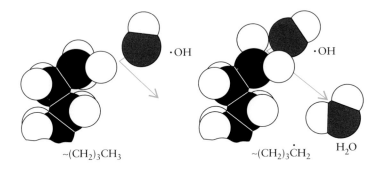

~(CH$_2$)$_3$CH$_3$ ·OH   ~(CH$_2$)$_3$ĊH$_2$ ·OH   H$_2$O

When an OH radical strikes a hydrocarbon molecule with a low impact velocity, no reaction occurs (left); however, if the energy of impact is above a threshold value, the atoms are loosened in the region of impact, and the OH radical may succeed in carrying off an atom from the hydrocarbon molecule.

However, if the OH radical smashes into the hydrocarbon with a shattering impact, then a hydrogen atom may be dislodged. For a fraction of a second the atoms form a loosely bonded cluster, then the hydrogen atom is dislodged, and it may be carried off. That is, if the energy of impact exceeds a certain minimum threshold value, which is called the "activation energy" of the reaction, the encounter will result in reaction.

The activation energy tells us the minimum energy that molecules need if they are to be able to react. However, we can construct a more detailed description of the energy changes that occur as the reaction takes place. Initially, a stationary, nonvibrating OH radical and a stationary, nonvibrating hydrocarbon molecule have a certain energy, which we mark on the illustration by the initial height of the curve. Since neither species is moving,

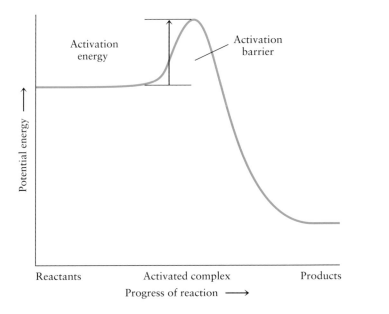

The reaction profile for the attack of an OH radical on a hydrocarbon molecule and the extraction of a hydrogen atom. The height of the barrier between reactants and products determines the rate at which the reaction can occur.

the energy is "potential energy," and is the energy stored in their chemical bonds. The fully formed stationary and nonvibrating product molecules—the H_2O molecule and the hydrocarbon radical—have a lower energy on account of their new bonding arrangement. As the reaction progresses, an OH radical approaches the hydrocarbon; the radical forms a bond to a hydrogen atom in the hydrocarbon and begins to draw it away from its carbon anchor; finally the new H—OH bond is fully formed and the new H_2O molecule escapes. The total potential energy of the system varies over the course of these events, as shown by the curve in the illustration. At short separations of the OH radical and the hydrocarbon, a new O—H bond is starting to form and a C—H bond is stretching as the hydrogen atom starts to ease out of its parent molecule. Because energy is needed to stretch the disturbed C—H bond, the overall potential energy is higher at this stage than initially. It rises to a maximum when the stress of disruption is greatest, and then falls as the newly created H_2O molecule moves to greater distances and the new O—H bond can relax to its final length.

The potential energy curve we have constructed is the "reaction profile." The maximum point of the curve corresponds to a more or less well defined cluster of atoms called the "activated complex," which is created for a brief moment by the encounter of the reactants. It is from this cluster of atoms that the products emerge. The height of the maximum relative to the reactants is the activation energy of the reaction; hence, it determines the characteristic temperature dependence of the reaction rate. When the OH radical and the hydrocarbon molecule are moving toward each other, they have a total energy that is the sum of the energy contributed by their motion, called "kinetic energy," and the energy contributed by their relative position, the potential energy. As the reactants approach, their kinetic energy falls as they slow down and their potential energy rises. Only if they still have a little kinetic energy at the top of the barrier will their total energy be greater than the activation energy, and they will be able to pass over the "activation barrier," the hump on the reaction profile, and emerge from the encounter as products.

The rates of the reactions in a candle flame depend on the proportion of collisions that occur with *at least* the activation energy of each reaction. This proportion is given by one of the most celebrated results of statistical thermodynamics, the "Boltzmann distribution," which was derived by the Austrian Ludwig Boltzmann in the late nineteenth century. The Boltzmann distribution can be used to deduce that the fraction of collisions that occur with at least a certain energy E_a in a gas at temperature T is $e^{-E_a/kT}$. The constant k is "Boltzmann's constant," a fundamental constant of nature that occurs widely in considerations of energy, entropy, and thermodynamics in general, and has a value of $k = 1.38 \times 10^{-23}$ J K^{-1} (joules per kelvin).

Ludwig Edward Boltzmann, 1844–1906.

CHAPTER FIVE

Since the rate of the reaction is proportional to the proportion of collisions with at least the energy E_a, we can immediately write

$$\text{Rate} \propto e^{-E_a/kT}$$

This is identical to the Arrhenius expression once E_a is identified with kT_a. Therefore, the existence of the activation temperature stems from the requirement that collisions must occur with at least a minimum energy (the activation energy) for the encounter to lead to products.

As the temperature approaches the activation temperature of the reaction, a higher proportion of the collisions occur with enough energy to react. When $T = T_a$, a fraction $1/e$ of the collisions (about 37 percent) are successful. Only when the temperature greatly exceeds the activation temperature is almost every collision successful, because then almost every collision occurs with an energy in excess of the activation energy. At temperatures well below the activation temperature, hardly any of the collisions occur with enough energy to result in reaction. Thus, except in the upper reaches of the atmosphere, atmospheric nitrogen does not react with atmospheric oxygen even though the two types of molecules undergo countless myriads of collisions: almost all the collisions are too feeble and the lower reaches of the atmosphere survive the potential conflagration.

The Boltzmann distribution is a result founded in chaos, for in his derivation of it Boltzmann assumed that the molecules of the system are able to take *at random* any energy that is consistent with some fixed, total value for the entire sample. That the Boltzmann distribution is rooted in chaos implies that the rates of reactions, just like the tendency of reactions to occur, are aspects of disorder. Although the direction of natural change is toward chaos, it is chaos that holds back systems from precipitate collapse by diluting the availability of energy: chaos is both the carrot and the cart of chemistry.

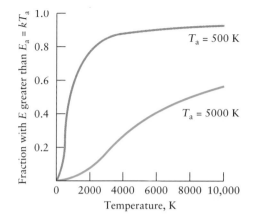

The fraction of collisions that occur with an energy of at least E_a increases with temperature. The higher the temperature, the greater the fraction of collisions that have enough energy to result in reaction.

REACTIVE ENCOUNTERS IN SOLUTION

Whereas the molecules in a gas can be thought of as traveling over considerable distances before smashing into each other, in a liquid the molecules are like a dense crowd of people: they jostle each other as they move, and each molecule can travel only an atom's width or so before colliding. We cannot think of a reaction in solution as occurring when molecules in flight smash together. However, we can still apply the concept of an activation energy: we can think of a reaction in solution as taking place as a result of the

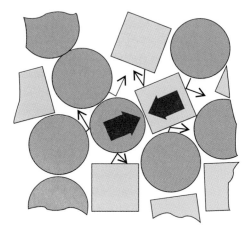

Two reactant molecules that lie next to each other may acquire sufficient energy to react if a fluctuation in the impacts of the surrounding molecules leaves them with excess energy.

accumulation of energy in a pair of reactant molecules that happen to be close together. A chance fluctuation in the impacts of a large number of molecules may supply a potentially reactive pair of molecules with the energy they need to react.

It turns out that the probability that a pair will attain at least the energy E_a from the fluctuation in the random storm they experience is also given by Boltzmann's exponential factor. Hence, reactions in solution—such as those that we stimulate in the kitchen and the ones we take for granted in our bodies—also typically show Arrhenius behavior. Such reactions are faster at high temperatures than at low because the local accumulations in energy are greater at high temperatures than at low. When we cook, we increase the violence and frequency of local fluctuations to speed the reactions on their way.

Actually, there is another dimension to reaction rates in solution, since some species react as soon as they meet without having to accumulate energy, yet still show Arrhenius behavior. For instance, although calcium ions and carbonate ions mass together as soon as they encounter each other, the reaction is not instantaneous because to meet, the ions—dressed in their cloaks of hydrating water molecules—need to worm their way past the intervening solvent molecules. Each hydrated ion must release itself from the weak attractions of the surrounding water molecules, then release itself from the neighborhood of another set of water molecules, and hence gradually migrate—the technical term is *diffuse*—on a random path through the solvent. Diffusion through a solvent is an "activated process" in the sense that each step can be accomplished only by ions that have more than a certain minimum energy, and so whether or not a particular ion can move at any instant is governed by the Boltzmann distribution.

It is common when discussing solution kinetics to distinguish two types of process, both of which exhibit Arrhenius behavior. In one, migration through the reaction mixture is easy, and the rate of reaction is determined by the probability that, having encountered each other, the molecules will accumulate enough energy to react successfully. In these "activation-controlled reactions," the activation energy is determined by the details of the energy requirement of the encounter pair; this is the case with the esterification reaction used to model candlewax formation. In the other type of reaction, which is called a "diffusion-controlled reaction," the activation energy of the encountering pair is so small that the rate of the reaction is determined by the rate at which the reactant molecules can wriggle through the solution. The precipitation of calcium carbonate that Faraday used to demonstrate the formation of carbon dioxide is an example of the latter type of reaction, for once the ions have met only sufficient energy is required to strip away the coating of hydrating molecules so that the naked ions can adhere together.

CHAPTER FIVE

DECELERATION AND ACCELERATION

Because the activation temperatures of most reactions are typically much higher than everyday temperatures, almost all the molecules in a sample are unavailable for reaction. The molecules that do temporarily have enough energy to react are the tiny proportion of adventurous spirits that live life dangerously—one collision, and they might become another substance. Reactions do not suddenly overwhelm us, even those that have a tendency to occur, because these currently adventurous spirits are only a very small proportion of the whole. If energy were more readily available for atoms to react, then all the rich variety of substances would long ago have sunk into the ultimate tar. If that risky spontaneous reaction that we know as combustion had only a tiny activation energy, then conflagration would have occurred as soon as a substance evolved. Indeed, without activation energies there could have been no life, for life depends on the judicious levering of selected species over energy barriers. If we want to make a reaction run faster, we need to increase the proportion of molecules that have an energy greater than the activation energy, perhaps by raising the temperature and elongating the tail of the Boltzmann distribution. At high temperatures, all molecules live dangerously.

The classic method of quenching a fire by pouring water makes use of the Boltzmann distribution and the activation energy of reactions. The evap-

One of the reasons why water is used to extinguish fire is that its vaporization is so endothermic that it lowers the temperature of the burning material so much that the species taking part in the conflagration do not have enough energy to cross the activation barriers of the combustion reactions.

oration of the water is strongly endothermic, and results in so great a reduction in temperature that the combustion reaction ceases. More modern versions of this highly effective but messy and sometimes inappropriate procedure include spraying the flame with aluminum hydroxide. The ensuing highly endothermic dehydration reaction

$$2Al(OH)_3 \longrightarrow Al_2O_3 + 3H_2O$$

effectively sucks the life out of the flame by cooling it below the temperature at which the chain reaction can propagate.

It is possible to achieve faster reaction without resorting to an increase in temperature; one way is use "catalysis" to hack out a new chemical pathway—a different series of chemical reactions that lead from the same starting materials to the products. A catalyst is a substance that facilitates a reaction but is not consumed in the process (the Chinese characters *tsoo mei* for "catalyst" translate appositely as "marriage broker"). Thus, the catalytic converters in automobiles speed the oxidation of unburned hydrocarbons to carbon dioxide but are not themselves consumed. A catalyst works by opening up a new reaction pathway that has a lower activation energy than the pathway the same reaction must follow in its absence.

Faraday was aware of the catalytic power of some metals (indeed, the term "catalysis" was coined in 1836 by the great Swedish chemist J. J. Berzelius, who dominated so much of chemical thought in the early nineteenth century). In his lectures Faraday demonstrated the catalytic power of the metal platinum when he showed the metal glowing brilliantly in a stream of hydrogen. This luminosity is a sign that a catalyst is at work, since the high temperature arises from the greatly accelerated combustion reaction. We now know that the catalytic ability of platinum stems from the ability of hydrogen to adhere to the surface of the metal as individual atoms. All oxygen needs to do in the combustion reaction is to pluck hydrogen atoms off the surface of the metal, and the contribution to the activation energy from the disruption of the H—H bonds in the gas is circumvented.

Catalysts are one of chemistry's greatest contributions to modern life, for in their current highly developed form they are used widely throughout chemical industry to open reaction pathways that would otherwise be closed to reagents or simply too expensive to exploit. An everyday offshoot of the catalytic power of platinum is seen in the increasingly refined complexity of the catalytic converters in automobiles. These converters are intended to complete the combustion of hydrocarbons that began in the cylinder and may still be unfinished at the time of their emission through the exhaust. Such emission-control catalysts should ideally be able to handle the additional problem that, when the engine is hot, polluting oxides of nitrogen are formed that must be reduced to harmless nitrogen.

When a platinum wire is held in a stream of hydrogen gas, it glows with a brilliant incandescence as a result of its catalysis of the combustion of hydrogen.

The structure of the catalyst used in automobile exhausts to reduce polluting emissions. The small pictures on the right show the catalyst at increasing magnification (from top to bottom) under a scanning electron microscope.

A part of the challenge of developing an automobile catalyst system is to create one that can lower the activation energies of both oxidation and reduction reactions. The catalyst must also be able to operate under highly variable conditions (such as the flow of fuel and air), which fluctuate at the whim of the driver and the mercy of the traffic flow. Yet another component of the challenge is that, as emissions are greatest during the first few seconds after the engine is started, the catalyst should be effective when almost cold as well as at normal operating temperatures (which are close to 400°C).

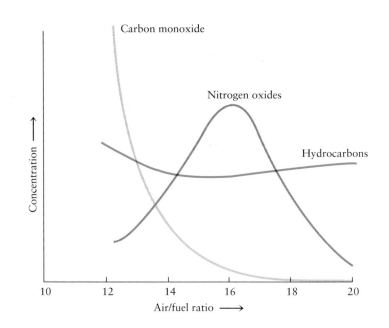

The concentration of pollutants formed in an internal combustion engine changes as the ratio of air and fuel is varied. When the conditions are fuel-rich, the combustion temperature is below the maximum and the formation of nitrogen oxides is low. The temperature is also low when the mixture is lean (excess air), and nitrogen oxide formation is again low.

There are three components to a typical modern catalytic converter: one to effect the reduction of nitrogen oxides, another to facilitate the oxidation of carbon monoxide and hydrocarbons, and the third to maintain the correct abundance of oxygen. In the first stage the nitrogen oxides are reduced using a platinum catalyst, which facilitates their decomposition into nitrogen and oxygen. In the next stage the carbon fragments are oxidized over a platinum/rhodium catalyst. Finally, the correct amount of oxygen is ensured by monitoring the amount of oxygen passing into the engine, and by incorporating into the catalyst a metal oxide that absorbs oxygen (by reacting with it to form a higher oxide) when the fuel mixture has too much oxygen and reverts to the lower oxide, releasing oxygen, when the mixture has too little.

THE EFFECT OF NUCLEAR MASS

Now we turn from the effect of temperature on reaction rates to the effect of atomic mass. Changing the atom of one element for the atom of a heavier element results in extensive changes to the electron distribution in the molecule. Thus the reactions of carbon disulfide, CS_2, are strikingly different from those of carbon dioxide because the electron distributions in the two molecules are very different. To show how mass itself affects a reaction rate, we must replace an atom of one element by an atom *of the same element* but a different mass.

Faraday would have been surprised to know that, contrary to John Dalton's original views on the nature of atoms, the atoms of an element are not all the same. All elements exist as various "isotopes," in which the electric charge of the nucleus is the same for all the isotopes of an element but the mass of the nucleus is different. A natural sample of hydrogen, for example, consists mostly of atoms that have a proton for a nucleus, but two atoms out of 10,000 will have a nucleus that consists of a proton and another particle, the electrically neutral neutron, and these two atoms will each be twice as heavy as normal hydrogen atoms. The heavy atoms are atoms of deuterium, D, an isotope of hydrogen. The substitution of an element by one of its isotopes, such as the replacement of hydrogen by deuterium, leaves the electron distribution in the molecules unchanged but results in the presence of a heavier atom.

It is generally true that when an atom is replaced by its heavier isotope, the reaction decelerates. In the extraction of a hydrogen atom from a hydrocarbon by an OH radical, for example, when the hydrogen of the hydrocarbon is replaced by deuterium, the C—D bond is cleaved at a noticeably slower rate than the original C—H bond. Note that we might intuitively expect the heavier atom to participate more slowly in reactions, just as anything massive has more inertia. However, the *reasons* for this behavior are quite unrelated to the classical inertia of a heavy object—they are entirely quantum mechanical.

The fact that the slowing of reactions by an increase of mass has a quantum mechanical origin introduces a general point that we shall increasingly elaborate in the following pages. That is, many of the characteristics of chemical reactions, particularly their rates and the steps by which they occur, are intrinsically quantum mechanical and not open to classical explanation. When chemists are carrying out what seem to be highly classical laboratory procedures, such as heating, mixing, stirring, and distilling, they are (often unwittingly) encouraging particular *quantum mechanical* processes to occur. Indeed, when Faraday was igniting his candle, developing his vision of the events in the flame, and carrying out his ostensibly simple demonstrations of reactions (and particularly when he was busily electrolyzing), he was unknowingly deploying quantum mechanics to achieve his ends. We shall see a little of this nonclassical quantum conjuring here, and much more of it in later chapters.

An isotope is able to exert an effect on reaction rates in two ways— through its effect on the quantum mechanical "zero-point energy" of a bond and through the quantum mechanical effect called "tunneling." We shall look at these two effects in turn.

The zero-point energy of a bond is the minimum vibrational energy that the bond may possess. According to quantum mechanics, it is not possible to eliminate all the vibrational energy of an oscillator—even in a bond's lowest possible energy state, the atoms fluctuate around a mean separation and cannot be stilled completely. One way of understanding the existence of the zero point vibrational energy is to consider the consequences of the uncertainty principle, which tells us that a particle cannot simultaneously have both a well-defined position and a well-defined momentum. Therefore, an atom in a bond cannot be still at a given position, for stillness implies zero momentum. The motion forced on an atom by the uncertainty principle gives it energy, its zero-point energy, and it follows that the minimum energy of the OH radical and the hydrocarbon molecule is slightly higher than the reaction profile would suggest.

The zero-point energy of a bond depends on two characteristics. One is the stiffness of the bond: the more rigidly the atoms are held, the greater the zero-point energy. This is also a quantum mechanical effect that we can interpret in terms of the uncertainty principle, for a stiff bond is strongly confining, which implies that the position is reasonably certain, and therefore that the uncertainty in the momentum in high and the zero-point energy is correspondingly large. It follows that the zero-point energy in the acti-

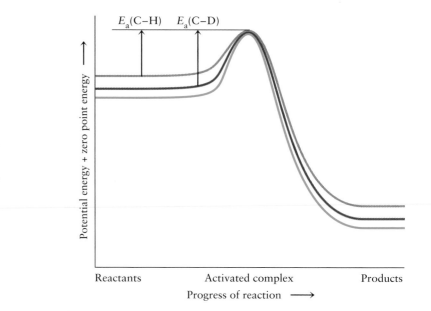

The zero-point energy of the reactants has the effect of raising the minimum energy they can possess above the line that depicts the reaction profile. The elevation of energy is greater for bonds containing hydrogen than for those containing deuterium, and is also greater for the stiffer bonds of the reactants than for the looser bonds of the activated complex. The outcome is a lowering of the activation energy for the cleavage of C—H bonds relative to that for C—D bonds.

CHAPTER FIVE

vated complex, where the atoms form a loose cluster, is much smaller than in the reactant molecules, where the atoms are held by normal, stiff chemical bonds. As a result, the minimum energy that the activated complex has is almost the same as the potential energy, as shown in the illustration.

The next crucial point is that the zero-point energy *decreases* with increasing mass of the vibrating atoms: as the mass increases, so the behavior becomes more classical, and consequently the zero-point energy approaches zero. That being so, when we replace a hydrogen atom by a deuterium atom, we can expect the zero-point energy to be reduced. It follows that the minimum energy of the reactants is greater for the reaction of a C—H bond than for the same reaction of a C—D bond. Since the energy of the activated complex is barely affected by a change in mass, the net effect is that the activation energy is higher for the deuterated species. Since for these *quantum mechanical* reasons the activation energy for the cleavage of a C—D bond is higher than that of a C—H bond, the reaction is slower.

Tunnelling is the ability of a particle to penetrate into—and through— classically forbidden regions. According to classical mechanics, if the reactants approach the barrier at the peak of the reaction profile with less kinetic energy than is needed to cross over it, then they will certainly be reflected back. In practice that means that an OH radical might approach a hydrocarbon molecule, but be unable to extract a hydrogen atom from it because it has insufficient kinetic energy to form the activated complex. However, when we turn to see what quantum mechanics tells us, we find that it allows reactants to cross the barrier and form products even though their kinetic energy is less than the activation energy. That is, the barrier is quantum mechanically porous.

To see the origin of the quantum porosity, we take up a point made in Chapter 2, where we saw that the locations of electrons in atoms are described by orbitals. In general we speak of an orbital as the "wavefunction" of a particle, a wavelike distribution that behaves smoothly everywhere and which determines where the particle is likely to be found. Wavefunctions— of electrons, protons, or any type of particle—do not terminate abruptly, but decay smoothly toward zero.

The illustration is a depiction of a wavefunction for a single particle, such as a hydrogen atom, approaching a barrier that should be impassable for a particle with its kinetic energy. At the barrier, the particle's wavefunction separates into two components. One component represents reflection back from the barrier. The other component continues inside the barrier, but it gets smaller the deeper the penetration. If the barrier is not very wide, the wavefunction might not have reached zero before the region of low potential energy beyond the barrier has been reached, and after passing the barrier the wavefunction represents a particle that has penetrated right

The wavefunction for a particle traveling from the left and striking a barrier. The wavefunction has been drawn in a somewhat unconventional manner to represent the fact that a traveling particle is represented by a complex wavefunction (one of the form a + ib, *where* i = $\sqrt{-1}$: *the real part of the wavefunction (the one normally shown, like that on p. 20) is the projection of the spiral on a vertical plane, and the imaginary component is the projection of the spiral on the horizontal plane. The direction of propagation of the particle is given by the right-handed screw sense of the complex wavefunction.*

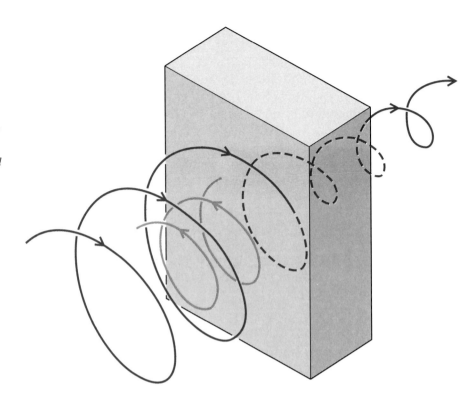

through and is traveling to the right. In classical terms, tunneling can be compared to a ball that has been thrown against a wall, and is then found beyond the wall even though it was thrown with an energy that was too low for it to break through.

The lighter the particle, the greater the probability that it will travel through a given barrier. Hence, the most likely particle to show signs of penetration through barriers is the electron (which has a very low mass). When Faraday was carrying out his electrolysis experiments, he was unwittingly making use of the ability of electrons to tunnel between his electrodes and the species in solution: without their ability to tunnel, electrons would be confined much more strictly within the metal electrodes and electrolysis would be impractically slow. For our present purposes though, we are concentrating on nuclear isotope effects, and so we focus on the proton and its heavier analog, the deuteron (the nucleus of a deuterium atom). As a consequence of tunneling, a C—H bond can fracture even though there is insufficient energy available to supply the activation energy of the reaction: the hydrogen atom can escape *through* the barrier without needing to accumulate enough energy to clamber over it.

It would probably be very difficult to detect the effects of tunneling in the reactions taking place in the candle flame, but other reactions do reveal them. Tunneling in chemistry is detected by observing a deviation of the rate from an Arrhenius-like temperature dependence, and in particular by observing that the rate of a reaction at low temperature is greater than would be expected by extrapolation of the high-temperature rates. The reaction appears to proceed with a lower activation energy at low temperature because the reactants do not need to acquire the full amount of energy to cross the barrier that separates them from the products. An example is the ability of a methyl radical, $\cdot CH_3$, to extract a hydrogen atom from a neighboring molecule in solid acetonitrile, CH_3CN, at low temperatures (80 K) in the reaction

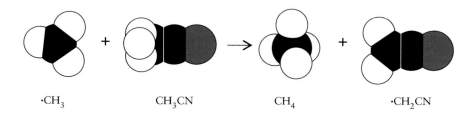

| $\cdot CH_3$ | CH_3CN | CH_4 | $\cdot CH_2CN$ |

On the basis of the activation energy measured at ordinary temperatures in the gas phase, no reaction at all would have been expected between the two species at such low temperatures. However, since a reaction does take place, it is inferred that tunneling is occurring. Because heavy particles are less likely to tunnel than light particles, another sign of tunneling is the great reduction in rate of a reaction when the mass of a reactant is increased. The rate of the reaction shown above is dramatically slowed—by a factor of 28,000—when CD_3CN is used in place of CH_3CN.

Although tunneling effects are most visible for protons, there is some evidence that larger particles—indeed, entire molecules—can tunnel through some barriers and lead to reaction when, on classical grounds, none would be expected. One example has been observed in the polymerization of formaldehyde, the linking together of formaldehyde molecules, CH_2O, to form a chain. One propagation step in the polymerization reaction is

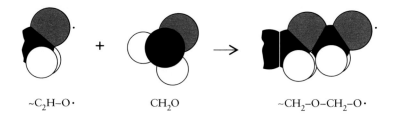

| ~C_2H–O$\cdot$ | CH_2O | ~CH_2–O–CH_2–O$\cdot$ |

The time it takes for one CH_2O unit to be attached to the chain can be measured, and is found to increase as the temperature is reduced, showing Arrhenius behavior in the range 150 K down to 80 K. At the latter temperature, the linking time is about 10 μs (10 millionths of a second). However, below 80 K the rate becomes distinctly non-Arrhenius, and at temperatures when extrapolation of the higher temperature rate would lead us to expect a linking time of about 10^{30} years, the chain continues to grow in steps that take only 1 millisecond for each link to be formed. Moreover, substitution of deuterium leads to almost *no* discernible change in rate. That this absence of an effect, paradoxically, is evidence for tunneling can be understood if it is the entire CH_2O molecule that tunnels through an activation barrier, for then the replacement of two hydrogen atoms by deuterium results in less than a 10 percent increase in mass. Such whole-molecule tunneling, it has been speculated, may be important in the cold (10–20 K) conditions of interstellar space, where simple molecules like formaldehyde and hydrogen cyanide on dust grains and larger bodies may be able to react even though they are frozen out of classical reactions.

REACTION BY REDEPLOYMENT

Tunneling is one way of beating the activation barrier, but it is limited to the reactions of atoms and of tiny fragments of molecules. A more sophisticated way of beating the barrier becomes apparent if we think about the origin of the Arrhenius expression. As we have seen, it stems from the Boltzmann distribution, which in turn is based on the presumption that there is a completely random distribution of energy over the available states. These energy states correspond to the various modes of motion present in the vibrating and traveling molecules. However, suppose we organize the energy available in the reactants so that it is not randomly distributed over all the available modes of motion, but is channeled into specific modes. Although the activation barrier will still have its same height, energy may be accumulated in the mode of motion that is most conducive to reaction.

We need to know what state of excitation is best for a particular reaction. Is the probability of reaction greater if the energy of collision of two molecules is entirely the kinetic energy of flight through space, or would it be better for one of the molecules to be traveling more slowly but vibrating as well (and have the same total energy)? If the former, one straightforward way of achieving the appropriate energy state is to shoot molecules at each other in a molecular beam, the chemists' version of the particle accelerators that physicists use to study their elementary particles. If, on the other hand, the chances of reaction increase when a reactant molecule is vibrating more energetically, it can be made to do so through exposure to infrared radiation. In many cases, such a redeployment of the total energy from an external mode of motion (flight) into an internal mode of motion (vibration) can speed a reaction dramatically.

To see why redeployment of energy may be an advantage, we need to elaborate our concept of reaction profile. The conceptual step we need to take is to realize that the reaction profile is a two-dimensional cut through a reaction hypersurface. We shall first unfold what this means and then show how it is relevant.

We shall go on plumbing the depths of the reaction of an OH radical with a hydrocarbon molecule, in which the radical plucks a hydrogen atom from the molecule and carries it off. A complete reaction profile of even this simple reaction would have to be drawn in a hyperspace of several dimensions. Thus, suppose we wanted to express the total potential energy of the reaction at any point in its course in terms of the location of the three species, the OH radical, the H atom it is attacking, and the C atom to which that atom is attached. To specify the location of three particles, we need to specify nine coordinates (three for each one), and so the potential energy is a function of nine variables. It follows that we need ten dimensions to portray the potential energy graphically.

Fortunately, though, the essentials of the problem can be represented in four dimensions. It follows from the symmetry of the system that there are in fact only four relevant coordinates, the separations of the atoms, specifically $R_{OH,H}$, $R_{H,C}$, and $R_{OH,C}$. So, to show the potential energy we actually need only three dimensions for the coordinates, and one for the value of the potential energy.

Even four dimensions is too much for ready visualization (computers, of course, will happily handle any number of dimensions as an exercise in bookkeeping, so for computational purposes high dimensionality is not a

The distances in blue define the positions of the atoms in the reaction in which an OH radical (treated as a single "atom") extracts a hydrogen atom from a hydrocarbon molecule. If the atoms are supposed to lie in a straight line (right), only two distances are needed to specify the locations.

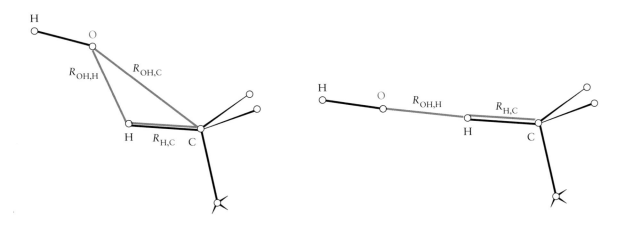

conceptual problem). Therefore, we shall cut away one degree of freedom by considering only reactions in which the particles lie in a straight line, and the OH radical approaches and departs along the direction of the CH bond. For this arrangement, $R_{OH,C} = R_{OH,H} + R_{H,C}$. Now we can express the potential energy in three-dimensional space. However, we should not forget that the three-dimensional surface is really a slice through a four-dimensional hypersurface, and that motion off the diagram in a direction perpendicular to the axes shown corresponds to a motion of the particles that destroys their collinearity.

The three-dimensional potential energy surface shown in the illustration has all the features we should expect. Thus, when the OH radical is very far from the hydrocarbon molecule, the potential energy varies as the C—H bond length changes exactly as we would expect for a diatomic molecule, as shown by the green curve in the illustration. The same is true at the end of the reaction, when the new HO—H molecule is far from the carbon atom in its parent hydrocarbon; now the potential behaves like that of an isolated HO—H bond (the yellow curve).

The crucial attribute of the potential energy surface can be appreciated by following the path of the blue curve through the valley in the center of the potential energy surface. The interesting features of the path are hidden by the hill in front, but are visible in the contour diagram below: the valley inclines upward until it reaches a pass—a "saddle point"—connecting the valley on the left with the valley on the right, then slopes down the valley on the other side. This is the path of least potential energy that can transform reactants into products. The upward slope of the valley at the beginning of the path reflects the increase in energy that stems from the stretching of the C—H bond, but this increase is partly offset by the incipient bond that is formed between HO and H as the OH radical approaches the hydrocarbon. The reactants are able to follow this path when the two bond lengths have been carefully adjusted to achieve the lowest energy at all stages from reactants to products. The path of minimum potential energy is essentially the reaction profile that we have considered so far.

We see that the surface is unsymmetrical in the sense that the saddle point—the point of highest energy on the valley between reactants and products—is late on the path. (I am guessing that the high point is late on the path and not early.) Physically this means that the highest energy on the reaction profile occurs when the original C—H bond has lengthened and the OH radical is very close to the H atom that it is going to abduct. This could also have been seen in the reaction profile we looked at previously, but in three dimensions a new feature comes to light: the shape of the valleys close to the saddle point.

The OH radical approaching the hydrocarbon molecule on trajectory T is targeted on a CH group that has little vibrational motion. On this trajec-

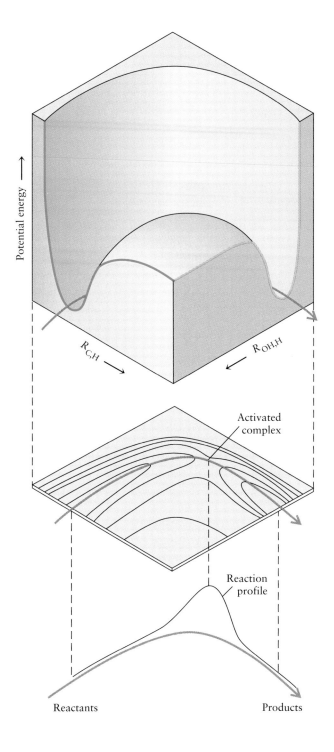

Potential energy →

$R_{C,H}$ →

← $R_{OH,H}$

Activated complex

Reaction profile

Reactants

Products

The potential energy surface for the approach of an OH radical to a C—H bond of a hydrocarbon molecule; beneath it is shown a representation in terms of contours of constant potential energy. The reaction profile is the path of minimum potential energy up the floor of the reactant valley, across the pass at the activated complex, and down the foot of the product valley.

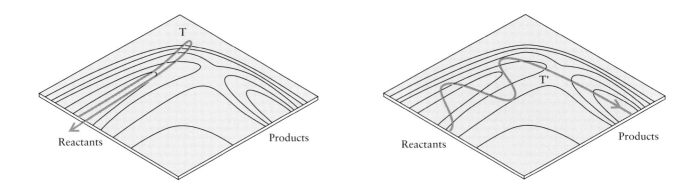

Reactants   Products   Reactants   Products

On the trajectory marked T (left) the kinetic energy is almost entirely due to the translational motion of flight: the C—H bond length is almost constant as OH approaches. On trajectory T' (right) the C—H bond is vibrating as OH approaches, and so the kinetic energy has two components; moreover, the H atom periodically reaches out to the approaching OH radical and shrinks away.

tory, the C—H bond is unyielding: it does not stretch, and the path stays to the left of the saddle point. The potential energy rises as the radical approaches the unyielding CH group and the path takes the system up the opposing face of the valley. It is easy to believe that, even though the kinetic energy is high enough for reaction, the system will be reflected back, like a ball thrown at a wall, and that the OH radical will leave the hydrocarbon molecule intact.

Now consider trajectory T' in which some of the same total kinetic energy is present as vibrational motion of the C—H bond. As the OH radical approaches, the C—H bond stretches and contracts, and the trajectory weaves from side to side up the valley as it approaches the saddle point. The saddle point represents the arrangement in which the HO—H distance is quite short and the C—H distance is long—a stretched and weakened C—H bond. The weaving motion of the trajectory may be sufficient to tip the system round the corner of the valley and then on to products. The HO radical runs off with the H atom, and maintains a nearly constant HO—H distance: that is, on this path the molecule is formed in a vibrationally inactive state.

It follows from this analysis that the reaction—and all those with similar potential surfaces—should go more efficiently if the excess energy is present as a vibration of the hydrocarbon. An example provides a glimmer of how such manipulations of the energy requirements of reactions might be used in the future as laser radiation—which can be used to excite specific vibrations of molecules—becomes cheaper. In a conventional reaction environment, boron trichloride, BCl_3, reacts with benzene, C_6H_6, to form $C_6H_5BCl_2$ in the presence of a palladium catalyst, but only at temperatures in excess of 600°C. However, the products will form at room temperature without a catalyst if the reactants (more specifically, the intermediate products that are formed in the course of the reaction) are vibrationally excited with 10.6-μm infrared radiation from a carbon dioxide laser.

DETAILED DYNAMICS

The potential energy surfaces of reactions can be used in combination with classical mechanics to develop insight of extraordinary detail into chemical reactions. It has long been possible to calculate the detailed motion of atoms at the climax of reaction; it is now possible to observe that motion as well. The calculations reveal how brief that encounter is, and, for more complex reactions, how turbulent.

The illustration shows the result of such a calculation for the motion of the three atoms in a very primitive reaction in which a hydrogen atom collides with a hydrogen molecule, and plucks off one of the hydrogen atoms. The lines show very clearly the vibration of the original molecule as the incoming molecule approaches. The reaction itself—the switch of partners—takes place very rapidly, almost within the period of one molecular vibration. The new molecule shakes for a fraction of a moment, but then settles down into steady harmonic vibration as the expelled atom departs.

The other illustration shows a much more complex sequence of events. In this reaction, four atoms exchange their partners:

$$KCl + NaBr \longrightarrow KBr + NaCl$$

The calculated trajectories show that the loose, turbulent cluster of four atoms survives for about 5 ps (1 ps = 10^{-12} s, one million-millionth of a second), which is time for the atoms to oscillate about fifteen times before falling apart with their new partners.

These calculations, of course, use classical mechanics, not quantum mechanics. If we were to do a full quantum mechanical calculation, the

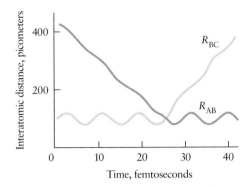

The calculated trajectories for a reactive encounter between a hydrogen atom A and a vibrating H_2 molecule (BC) leading to the formation of an AB molecule. This rapid exchange of partners is an example of a "direct mode" reaction.

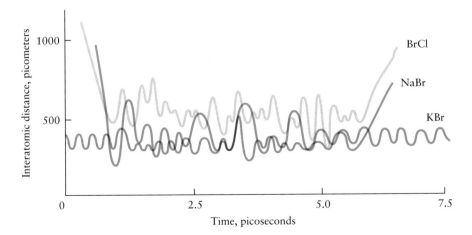

An example of the trajectories calculated for a "complex mode" reaction in which the collision complex has a long lifetime and executes complicated motions before the products emerge from it.

concept of trajectory would fade and be replaced by that of an unfolding wavefunction that initially represents reactants and finally products. Such calculations can be done, or at least quantum mechanical concepts and classical trajectories can be cobbled together to give an approximate picture of what is happening, and these calculations lead to quantitatively reliable predictions of reaction rates.

FEMTOCHEMICAL REACTIONS

Until very recently there were no direct spectroscopic observations on the motion of atoms such as those we have just described, for the cluster of atoms that marks the climax of the reaction survives for only a few picoseconds before it is frozen into products and separates into distinct entities. However, calculation can now be verified by experiment since the development of lasers that generate pulses of femtosecond duration (1 femtosecond is 10^{-15} s, one million-millionth of a millisecond), for these lasers have opened up opportunities to see even reactive events as though they were frozen in time.

In a typical experiment, a femtosecond pulse is used to excite a molecule until it dissociates into fragments; then a second femtosecond laser pulse is fired at a series of intervals after the dissociating pulse. The second pulse is set at a frequency that will be absorbed by one of the free fragmentation products, so its absorption is a measure of the abundance of product. The current crop of results do not include the reactions that occur in the candle flame, so we need to digress slightly into other reactions; however, much the same results are likely to apply to our reactions too.

The laser system used to resolve the dynamics of reactions on a femtosecond timescale. The laser generates a pump pulse and a probe pulse. They leave the laser at the same time, but the probe pulse is diverted along a slightly longer path to introduce a time delay of a few femtoseconds between the two pulses (light travels only 0.0003 mm in 1 femtosecond). The pump pulse initiates a chemical reaction, and the probe pulse arrives a few femtoseconds later and causes the molecule to emit light, which is detected and analyzed.

CHAPTER FIVE

We can gain some sense of the progress that has been made in the study of the intimate mechanism of chemical reactions by considering the observations on the ion pair Na^+I^-, where I^- is an iodide ion. The dissociation of this pair has been studied by exciting it with a femtosecond pulse to an excited state that can be interpreted as a covalently bonded NaI molecule. The second probe pulse tracks the decay of the NaI molecule by examining the system at an absorption frequency either of the free Na atom or of the Na atom when it is a part of the complex.

A typical set of results is shown in the illustration. The variation in the absorption intensity of the bound sodium atom shows up as a series of pulses that recur about every 1 ps. The absorption frequency of the bound Na atom depends on the Na—I distance, so the variation indicates that the complex oscillates with a period of 1 ps. An absorption is obtained each time the vibration of the cluster returns it to a separation corresponding to the bound atom. The decline in intensity shows the rate at which the NaI molecule dissociates as the two atoms swing away from each other. The molecule has a chance to dissociate when the vibration stretches its bond, and as successive opportunities for escape are presented, the abundance of surviving molecules decreases. The absorption by the free sodium atom also grows in an oscillating manner as more atoms make their escape on each successive swing, and the increase in their abundance shows the same variation with time. These observations let us conclude that the period of the oscillation in NaI is 1.25 ps and that the molecule survives for about ten oscillations. The oscillation frequency of NaBr is similar, but the molecule barely survives one oscillation: as soon as the atoms swing apart, they escape from the bond that held them.

The concept of "molecular structure" did not emerge until 1858, with a paper by the Scottish chemist A. S. Couper, which gave the first representation in print of a structural formula, and Faraday might well have been familiar with the ideas of molecular structure that emerged during the last decade of his life. He would certainly have been delighted to know that current experimental techniques are able to reveal the most intimate details of the changes in shape that molecules can undergo and to let us see, deep down, what determines the rates at which atoms can exchange their partners, achieve liberty, and give rise to new substances.

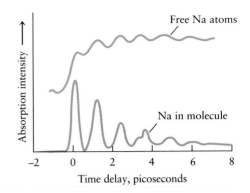

A femtosecond spectroscopic observation on the reaction in which an NaI molecule separates into Na and I atoms. The lower graph is the absorption of the complex and the upper graph the absorption of the free Na atoms.

CONCENTRATION, OSCILLATION, AND THE EMERGENCE OF FORM

6

*The pattern of a peacock feather is
determined genetically and expressed by
periodic chemical reactions that determine
the structure of the feather. Reactions that
oscillate in time or space can be described
by the differential equations that are
described in this chapter.*

*For the sake, therefore, of a little regularity, and to
simplify the matter, I shall make a quiet flame, for who
can study a subject when there are difficulties in the way
not belonging to it?*

Michael Faraday, Lecture 1

W

e have seen that the rate of a chemical reaction depends on the temperature at which it is carried out. The rate also depends on the concentrations of the species participating in the reaction. Faraday would have known that intuitively, and he might also have been aware of the precise quantitative relationships, for the dependence of certain reaction rates on concentrations had been established experimentally in 1850 by the German chemist Ludwig Wilhelmy working in Heidelberg. However, the systematic study of reaction rates, and their discussion in terms of the differential equations known as "rate laws," came at the very evening of Faraday's life, in 1865, when the Oxford chemist Vernon Harcourt and the Oxford mathematician William Esson pooled their complementary talents and established the formulation chemists still use today.

Faraday might have had little taste for such mathematicizing of so experimental a science as chemistry, but had he been both alive and receptive, he would have come to see that rate laws are a window on to the events that take place as molecules react, and he would have marveled at the insight that such a formal, quantitative approach to reactions could yield. He would also have marveled at the consequences of the marriage of mathematics and chemistry, for through that union the abstract implications of equations are endowed with a substantive corporality: the patterns of mathematics are translated into patterns traced by matter in space and time through the changing concentrations of reactants and products. He might also have taken a wry delight from the current status of the modern descendants of Esson's equations, for now the ability to predict the course of reactions seems to have slipped from chemists' grasp as smooth evolution of products has given way—in the theater of current interest—to chaotic generation and decay.

In this chapter we span the history of chemical kinetics, from Esson's original formulation to the cumulation of his equations, the onset of chaos.

In particular, we shall see what insights into the events that accompany reactions can be inferred from the concentration dependence of their rates. I want to convey the sense of the explosive advance in our ability to explore the rates of chemical reactions and their consequences that has occurred since Faraday's time, and particularly the new power the computer has placed in our hands to analyze the progress of reactions.

REACTION RATES

The rate of a chemical reaction is the change in concentration of a species per unit time. In some reactions the concentrations of the starting materials change very rapidly with time. An explosion is an example of a fast reaction, as are many combustions. On the other hand, the reactions that lead to rancidity and aging are mercifully slow.

An important point to note from the outset is that there is no such thing as a single "rate" of a reaction. We cannot in general speak of *the* rate of a reaction throughout its course any more than we can speak of *the* speed of a car throughout its journey. However, unlike a journey, where the speed of the car varies at the whim of the driver and the adventitious circumstances of the highway, the rates of chemical reactions vary much more systematically and can be expressed in terms of a mathematical expression called the "rate law" of the reaction.

An example of a rate law is the one for the decomposition of the gas ethane, C_2H_6, when it is strongly heated in the absence of air: we can think of the reaction as modeling some of the processes that rip longer hydrocarbon molecules apart as they vaporize into the heat of the candle flame on their route to becoming carbon dioxide. The reaction results in the formation of a mixture of ethylene (C_2H_4), methane (CH_4), hydrogen, and some butane (C_4H_{10}) as hydrogen atoms are ejected from the molecule and the fragments combine. It is found that, under certain conditions, the rate of decomposition of ethane is proportional to its concentration:

$$\text{Rate of decomposition} = k[C_2H_6]$$

In this expression, the "rate constant" k is characteristic of the reaction and is determined by experiment. The rate law shows that the rate of the reaction decreases—the reaction decelerates—as the concentration of ethane decreases. Some rate laws are more complicated than this one. For example, a reaction that occurs in the upper atmosphere is the decomposition of

The reaction that occurs when liquid bromine is poured onto solid phosphorus is vigorous, and the amounts of the reactants change rapidly as the product (phosphorus bromide) is formed. This is an example of a reasonably fast reaction. Some reactions, however, are very slow and occur with a barely discernable rate at room temperature.

ozone, O_3, into molecular oxygen, O_2, and a simplified version of its rate law is

$$\text{Rate of decomposition} = \frac{k[O_3]^2}{[O_2]}$$

This rate law shows, among other features, that the reaction is slower if it is carried out in a region where the concentration of *product* is high—the formation of oxygen acts as a brake on the reaction. How can it be that the presence of product impedes the rate at which product is formed?

There must be a *reason* why a reaction obeys a particular rate law—why, for instance, the formation of O_2 interferes with its own production—and looking for that reason is a very important technique for determining the "mechanism" of the reaction, the series of steps by which the reaction takes place. The determination of the rate law of a reaction is typically the first (and sometimes the only) means a chemist has for the determination of the reaction mechanism.

Although a rate law is a window on a reaction mechanism, it is a cloudy window, and chemists have learned to be cautious in the interpretation of what they see through it, for a number of different mechanisms can lead to the same rate law. Although a rate law is very helpful, it alone is only *indicative* of a mechanism: it is circumstantial evidence, not definitive evidence. For this reason, chemists speak of the "currently accepted" mechanism of a reaction rather than "its" mechanism. They cautiously accept that the "proof" of a mechanism in chemical kinetics is more like the proof of a case in a court of law: the evidence is accumulative, but rarely absolute. This attitude is changing as techniques are developed that explore reactions ever more intimately, and in many cases we can be very confident indeed about a kinetic analysis. However, it would be as well to bear in mind that the mechanisms we shall describe are not frozen—in a few months time, or even already yesterday, one or more of them will be shown to be erroneous and to be ripe for replacement by superior but perhaps still temporary alternatives.

SOME TYPICAL RATE LAWS

Many different reactions have rate laws of the same general form (but with different values of k), and so their rates vary with concentration in the same way. This observation lets us classify reactions according to their rate laws. Thus, we shall see that many reactions are "first-order reactions" (in a sense to be defined) and many are "second-order reactions," and so on. It is often

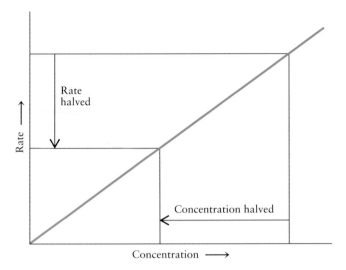

The rate of a first-order reaction is proportional to the concentration of the reactant. Thus, when the decomposition has reduced the concentration of ethane by one half, the rate of decay will have slowed to half its initial value.

very useful to know that a reaction belongs to one of the classes of reactions because predictions can then be made about its behavior from a knowledge of the general properties of reactions of its class.

The rate law for the decomposition of ethane given above is an example of a "first-order rate law," one in which the rate is proportional to the first power of the concentration of the substance. Other first-order decomposition reactions occur at different rates, and that difference is carried in the value of k: the greater the value of k, the faster the decay (for a given concentration of the decomposing species).

The principal characteristics of all first-order reactions is that the concentration of the reactant decays *exponentially* (as e^{-x}) with time, starting with the initial concentration of the reactant at time zero and gliding down

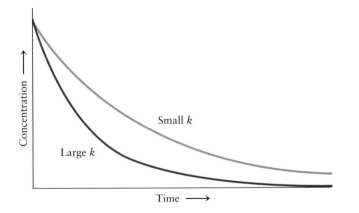

The concentration of a reactant decays exponentially toward zero if the reaction is first-order. The concentration decays more rapidly the higher the value of the rate constant.

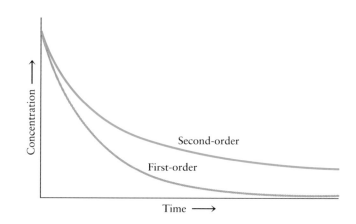

A first-order reaction (orange) and a second-order reaction (green) may have the same rates initially, but the second-order reaction slows markedly relative to the first-order reaction, and the reactants take much longer to disappear.

to zero concentration as time goes on. All exponential decays are the result of a rate law in which the rate of the events taking place is proportional to the concentration of the species that is being consumed. As the concentration approaches zero, so the rate at which it disappears also approaches zero. When there is only a very tiny amount of substance remaining, its disappearance is very slow.

In a "second-order reaction," the rate is proportional to the second power of the concentration of a species. For example, the rate at which methyl radicals, $\cdot CH_3$, combine in a gas to form ethane molecules is

$$\text{Rate of reaction} = k[\cdot CH_3]^2$$

Once again, k is a rate constant that varies from reaction to reaction. As the methyl radicals combine, the reaction decelerates sharply. If in a second-order reaction the concentration of a reactant falls to one-half its initial value, the rate falls to one-fourth its initial value. Under the same circumstance, the rate of a first-order reaction would have fallen to only one-half its initial value. That is, a second-order reaction decelerates toward zero more quickly than a first-order reaction with the same initial rate, and so the reactants vanish more slowly. In general, the higher the order of a reaction, the more responsive is its rate to changes in the concentration.

There is a common kind of hybrid reaction that is first-order in two different species and second-order overall. An example is the reaction originally studied by Wilhelmy and which marks the birth of quantitative chemical kinetics. In this reaction, sucrose (cane sugar), a molecule composed of a glucose and a fructose unit, decomposes into its components in dilute acidic solution:

$$\text{Sucrose } (aq) + H_2O(l) \longrightarrow \text{glucose} + \text{fructose}$$

Its rate law is

$$\text{Rate} = k[\text{sucrose}][\text{H}^+]$$

The reaction is first-order in hydrogen ions (and hence proportional to the acidity of the solution, so the reaction proceeds reasonably rapidly in the acidic environment of the stomach) and first-order in the sucrose. It follows from the rate law that if the concentration of *either* species is halved, then the rate is halved. If the concentrations of *both* species are halved, the reaction behaves like a second-order reaction, and its rate is reduced to one-fourth.

An example of a third-order reaction is the one that contributes to the formation of photochemical smog. The hot exhaust outlets of automobiles stimulate the oxidation of nitrogen to nitric oxide, NO, which on contact with the air is oxidized in turn to the pungent, poisonous, brown gas nitrogen dioxide, NO_2:

$$2NO(g) + O_2(g) \longrightarrow 2NO_2(g) \qquad \text{Rate} = k[\text{NO}]^2[\text{O}_2]$$

Photochemical smog (here, in Saõ Paolo, Brazil) is brown partly on account of the particulate matter present, which in scattering light imparts a red-brown hue to it, and partly on account of the presence of the brown gas nitrogen dioxide, NO_2, shown here on the right formed when the colorless gas nitric acid, NO, reacts with the oxygen in the air.

ACCOUNTING FOR SIMPLE RATE LAWS

A rate law is a window on to reaction mechanism because it is shaped by the events that occur during a reaction, and if we can account for a rate law we may have identified those events. We shall look at two mechanisms that can account for the behavior of first-order and second-order rate laws. Note how we are being cautious: there are many mechanisms that lead to the two types of behavior. We shall consider the *simplest* of them.

Some first-order reactions are simply the decomposition of a molecule in a single step, when it shakes itself apart. A first-order reaction is obtained if in any interval of time there is an equal likelihood that the molecule will fall apart into products (see the illustration). For example, there may be a one-in-a-billion likelihood that a particular energetically excited formalde-hyde molecule, H_2CO, formed briefly in a flame, may discard a hydrogen

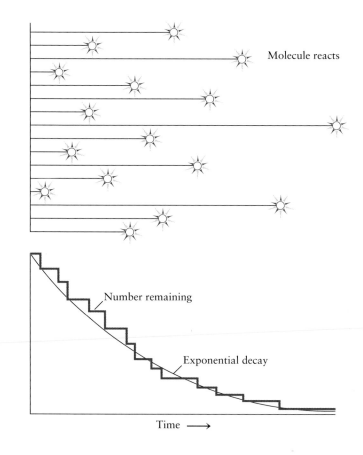

Molecule reacts

Number remaining

Exponential decay

Time $\longrightarrow$

Any given molecule in the system represented by the upper drawing has an equal chance of decaying in a given interval, and we expect to obtain exponential decay of the concentration of the substance. The lower curve shows the concentration remaining after each molecule reacts. When there are billions of molecules present, the concentration curve is a decaying exponential.

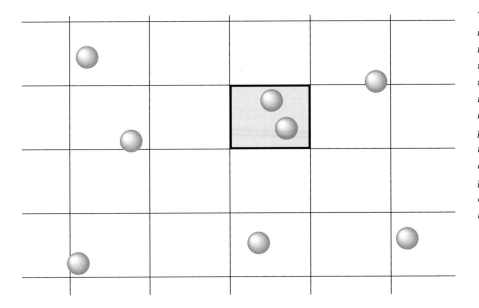

The probability of an encounter between two molecules is proportional to the probability that both will be found in the same small region (blue). The probability that any one molecule is in the shaded region is proportional to the concentration of molecules; hence, the probability that both are present is proportional to the square of the concentration. The probability that reaction occurs also depends on the energy that the two molecules possess when they meet: that point was discussed in Chapter 5 and is taken into account by the value of the proportionality constant k.

atom in any one-microsecond interval. We cannot predict which molecule will shrug off an atom, but we can be confident that one in a billion will do so in that interval. As time passes, more and more formaldehyde molecules will already have discarded a hydrogen atom, leaving fewer to do so. Since the number of molecules waiting on the brink of decomposition decreases with time, the rate of decomposition decreases as the concentration of formaldehyde lessens, and so we can anticipate that the rate law is

$$\text{Rate} = k[\text{H}_2\text{CO}]$$

The simplest version of a reaction that has a second-order rate law is one in which two species need to meet before they can react. For example, two methyl radicals need to meet somewhere before they can react to form an ethane molecule. The rate of this reaction is determined by the frequency at which the two species meet, which is proportional to their concentrations; hence we can write

$$\text{Rate} = k \times [\,\cdot\,\text{CH}_3] \times [\,\cdot\,\text{CH}_3] = k[\,\cdot\,\text{CH}_3]^2$$

with k a constant. This is exactly the form of a second-order rate law.

It is tempting to suppose that a third-order reaction (such as the oxidation of nitric oxide) is the outcome of a triple collision, since in that case the rate would be proportional to the product of the concentrations of the three simultaneously colliding species. However, triple collisions are generally very rare, and except in special cases (when they are the *only* way that a reaction can occur) can be ignored.

THE RATE-DETERMINING STEP

To account for third-order and other, more complex reactions, and to see how first-order and second-order reactions can arise when the reaction mechanisms are more complex than those we have discussed so far, we need to move up one notch of sophistication. In general, reactions occur in a series of steps—there are a myriad of steps that lead from candlewax to carbon dioxide, for example. All these individual steps occur at their characteristic rates, and the overall rate law is a combination of all their rates. In certain simple cases, however, one of the steps in the reaction mechanism dominates the rate, and then the overall rate law might be simple to establish.

A clue to how a single step can determine the reaction rate is found in a simple reaction with an intriguing property—it has a *constant* rate. A reaction with a rate that is independent of the concentration has a rate law with no concentration term appearing,

$$\text{Rate} = k$$

and is called a "zero-order" reaction because a concentration raised to the power zero is equal to 1 (that is $x^0 = 1$, so x does not appear in the rate law). An example of a zero-order reaction is the decomposition of ammonia into nitrogen and hydrogen on a hot tungsten wire, which proceeds at a steady rate until all the ammonia has decomposed, and then comes to a more or less abrupt halt.

How can a reaction continue at the same rate even though the reactant is being consumed—and then suddenly stop when the last remnants of the reactant have disappeared? The resolution of this puzzle will in fact prove to be an example of how different mechanisms can sneakily give rise to different rate laws. The key to the mechanism is the realization that rate of attachment of the NH_3 molecules to the surface of the hot metal is very fast. Once attached, an NH_3 molecule undergoes the complex series of reactions that results ultimately in the formation of N_2 and H_2 molecules. The rate of the reaction is determined by the rate at which these events occur, not the rate at which the NH_3 molecules attach to the surface. However, only the latter

NH$_3$ molecule

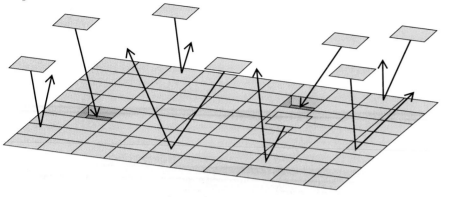

As soon as a vacant site appears on the surface of a catalyst, molecules from the storm of gas that is constantly raining down will seek it out, and a vacancy is never empty long.

step is governed by the concentration of the ammonia in the space above the metal. Since the attachment step plays no role in determining the rate of the reaction, the rate is independent of the concentration of the reactant—it is a zero-order reaction.

As in the decomposition of ammonia, the rate of a complex sequence of steps is frequently governed by the rate at which *one* of them—the "rate-determining step"—takes place. The rate-determining step in a reaction has an effect like the toll booths on a highway: although the traffic can approach the booths quite rapidly, the rate of passage through them governs the time taken for the journey. Now we can begin to see how it is possible to be misled by a rate law, for even though the reaction might have a very complicated mechanism, it may still turn out to be first- or second-order if that is the rate law of the rate-determining step.

An explicit example of a rate-determining step is found in the reaction proposed to account for the decomposition of ozone in the stratosphere (this reaction is crucial to the quantitative discussion of the disappearance of the ozone layer in polar regions, for it is one of the many processes that contributes to the steady-state condition of the atmosphere). The rate law was given on p. 132, where we saw that although the reaction rate is greater at high concentrations of ozone (as might be expected for a reaction in which two O$_3$ molecules collide and break apart into O$_2$ molecules), there is the peculiar feature that the rate is lower in regions of the atmosphere where the oxygen concentration is high.

The following (simplified) mechanism has been suggested to account for these findings. It is proposed that an energetically excited O$_3$ molecule first shakes itself apart in a first-order process:

$$O_3 \longrightarrow O_2 + O \qquad Rate = k_a[O_3] \qquad (a)$$

The O atoms produced in the decomposition of O_3 molecules are partly responsible for the formation of auroras: they contribute to their crimson and whitish-green colors.

The product of reaction (a) can go on to participate in one of two possible reactions. One possibility is for the reverse reaction to occur, in which an O atom attacks an O_2 molecule (not necessarily its earlier partner):

$$O + O_2 \longrightarrow O_3 \qquad \text{Rate} = k_b[O][O_2] \qquad \text{(b)}$$

Alternatively, the O atom may collide with an O_3 molecule, producing O_2 molecules:

$$O + O_3 \longrightarrow 2O_2 \qquad \text{Rate} = k_c[O][O_3] \qquad \text{(c)}$$

A proposed mechanism is not valid unless its rate law matches that already determined experimentally for the overall reaction, so a necessary (but not sufficient) test of this mechanism is to verify that the rate law it implies for the overall reaction is the one observed experimentally.

Measurements of the rates of the individual reactions show that step (c) is by far the slowest, and so we identify it as the rate-determining step. That being so, the rate of decomposition of ozone should be equal to the rate law for step (c). However, the rate law given there is not yet a true rate law for the overall reaction because it is not expressed solely in terms of the reactants (O_3) and products (O_2) but includes the concentration of a reaction intermediate (the O atom). The concentration of O atoms must be expressed in terms of those of the reactants and products. To do so, we recognize that reaction (a) and its reverse (b) are so fast (relative to the rate-determining step) that they quickly settle down into a dynamic equilibrium

$$O_3 \rightleftharpoons O_2 + O$$

in which the rates of the forward and reverse reactions are equal. That is, throughout the overall reaction we can assume that

$$\underset{\text{Rate of (a)}}{k_a[O_3]} = \underset{\text{Rate of (b)}}{k_b[O_2][O]}$$

This expression can be solved easily for [O], and the result substituted into the equation for the overall rate:

$$\text{Rate} = k_c[O_3][O] = \frac{k_a k_c[O_3]^2}{k_b[O_2]}$$

This expression is identical to the observed rate law if we identify the observed rate constant k with $k_a k_c / k_b$.

The mechanism reveals the physical reason why the reaction rate is slower in regions of the atmosphere where the concentration of oxygen is high. The O_2 molecules provide a route for the O atoms to re-create the O_3 molecules (in step b) instead of attacking another O_3 molecule to produce O_2 molecules (in step c). Thus, the higher the oxygen concentration, the more the O atoms are diverted from the formation of product.

ANTI-ARRHENIUS REACTIONS

Rate laws are sometimes used to uncover the reasons behind unusual behavior, such as we remarked upon in Chapter 5, when we mentioned in passing that some reactions show the opposite of Arrhenius behavior and go more slowly at higher temperatures. The reason can now be identified quite simply, for, as in the ozone decomposition reaction that we have just described, the observed rate constant of a reaction with a complex mechanism is a combination of the rate constants of the individual reaction steps. Even though all the rate constants increase with increasing temperature, their combination might not.

For instance, if in a three-step reaction like the ozone decomposition, k_b increases more rapidly than the product $k_a k_c$, the observed rate constant $k = k_a k_c / k_b$ will *decrease* as the temperature is raised, and the reaction will

This sequence of photographs shows the periodic color changes that occur in the Belousov-Zhabotinskii reaction, in which an organic compound (malonic acid) reacts with bromine formed by a complex series of reactions that involves bromide ions and bromate ions in the presence of cerium ions. The progress of the reaction is monitored by the color changes of a few drops of an organic dye, an indicator, that has been added to the reaction mixture but does not participate directly in the reaction.

on tigers and zebras and the spots and patches on leopards and giraffes, for as we shall see later, these markings are the outcome of spatially periodic chemical reactions in the (largely unstirred) skins of the animals concerned.

Oscillating reactions are much more than a laboratory curiosity. While they are known to occur in only a few cases in industrial processes, there are many examples in biochemical systems. Oscillating reactions, for example, maintain the rhythm of the heartbeat. They are also known to occur in the glycolytic cycle, a sequence of enzyme-catalyzed reactions that extract the energy of the glucose in food. In the steps of the cycle, one molecule of glucose is used to produce two molecules of energy-storing ATP. All the metabolites in the chain oscillate under some conditions, and do so with the same period but with different phases. Although we now recognize that oscillations occur in living systems, they were not initially recognized as oscillating chemical reactions, and the initial reports of oscillating reactions in a test tube were greeted with considerable skepticism. One of the first of these oscillating reactions to be reported and studied systematically was the "Belousov-Zhabotinskii (BZ) reaction." The reaction is named after the two Russian scientists who discovered and worked on the reaction (B. P. Belousov in 1951 and A. M. Zhabotinskii in 1964) and finally persuaded others that reactions could oscillate.

We shall find the complexities of the BZ reaction easier to grasp if we first describe an autocatalytic reaction of a particularly simple form that illustrates how oscillations may arise. The "Lotka-Volterra mechanism" is as follows:

$$A + X \longrightarrow 2X \qquad Rate = k[A][X] \qquad (a)$$

$$X + Y \longrightarrow 2Y \qquad Rate = k'[X][Y] \qquad (b)$$

$$Y \longrightarrow P \qquad Rate = k''[Y] \qquad (c)$$

The net effect of the reaction is the conversion of the reactant A to the product P. Steps (a) and (b) are autocatalytic. The actual chemical examples, including the BZ reaction, that have been discovered so far have a different mechanism, as we shall see, but the Lotka-Volterra mechanism will introduce us to concepts that are encountered in more realistic schemes too.

We shall consider the reaction as taking place inside a vessel into which the reactants are continuously flowing from outside and from which the products are continuously withdrawn. Such a "continuous stirred-tank reactor" (CSTR) maintains the reaction system far from equilibrium and allows the oscillations to persist indefinitely. The CSTR ensures that there is a constant supply of reactant A and a constant amount of P. As a result, the only concentrations that are changing are those of the intermediates X and Y. The human analog of continuous replenishment is the ingestion of food and excretion of its remains for the sustenance of the oscillatory reaction we know as the heartbeat and all the other chemical rhythms of the body. Admittedly, only a few people eat continuously, and hence few are strict analogs of a CSTR, but the process of digestion, as distinct from ingestion, smooths out the supply of metabolites to the various centers where they are needed, and so all of us are moderately good approximations to a CSTR.

We can anticipate some of the oscillatory features of the concentrations simply by inspecting the reaction mechanism. As soon as some X is present, its concentration surges upward by the autocatalytic step (a). However, as X grows in abundance, it reacts with Y to form more Y: this step is autocatalytic too, and there is a sudden surge of Y. That surge depletes the concentration of X, and that intermediate's concentration decreases rapidly as the concentration of Y grows. The concentration of Y cannot grow indefinitely, since it is removed by reaction (c), the rate of which increases as the concentration of Y increases. Hence, there comes a stage when the concentration of Y is so small that the autocatalytic reaction that produces it is effectively shut down, and the concentration of X has a chance to increase again. Now the cycle recommences, and so long as the supply of A is sustained, the concentration of the two intermediates continues to oscillate: as long as we eat, our heart will beat.

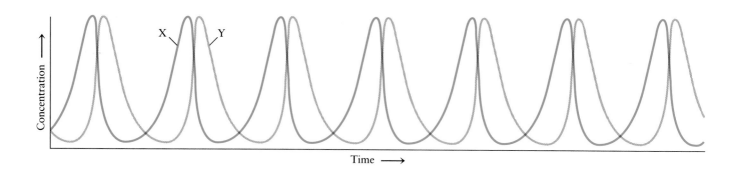

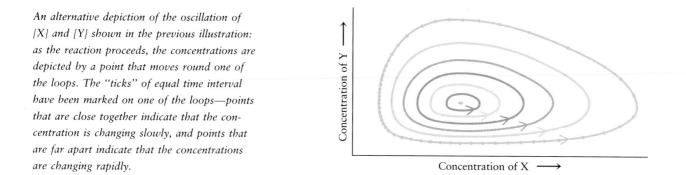

The periodic variation of the concentrations of the variables [X] and [Y] in the Lotka-Volterra mechanism: note how a surge in one is followed by a surge in another. The same type of equations have been used to study predator-prey relations in local ecosystems. Different frequencies and amplitudes of oscillation are obtained for different concentrations of A (and different rate constants).

When we write the rates of change of the concentrations of the intermediates, they form an interdependent pair of differential equations that show how the rates depend on the concentrations of A, X, and Y and the rate constants:

$$\frac{d[X]}{dt} = k[A][X] - k'[X][Y]$$

$$\frac{d[Y]}{dt} = k'[X][Y] - k''[Y]$$

Because we can vary the steady-state concentration of the reactant A in the reactor by varying the rate at which the CSTR is supplied with new A, we can use the concentration of A as a control to modify the reaction.

To find the time dependence of the concentrations of the intermediates, we need to solve the Lotka-Volterra equations numerically. The results can be depicted in two ways. One way is to plot [X] and [Y] against time. As the illustration shows, the two concentrations vary periodically with a rhythmic surge and sharp decline in each of their values, just as our qualitative analysis suggested. The same information can be displayed more succinctly by

An alternative depiction of the oscillation of [X] and [Y] shown in the previous illustration: as the reaction proceeds, the concentrations are depicted by a point that moves round one of the loops. The "ticks" of equal time interval have been marked on one of the loops—points that are close together indicate that the concentration is changing slowly, and points that are far apart indicate that the concentrations are changing rapidly.

CHAPTER SIX

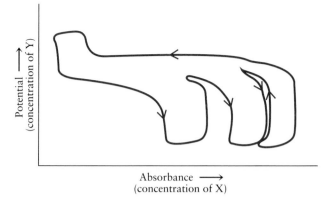

Potential ⟶
(concentration of Y)

Absorbance ⟶
(concentration of X)

A complex periodicity is displayed in a bromate-chlorite-iodide reaction. This pattern was obtained experimentally by monitoring the potential of an electrode immersed in the mixture and plotting it against the optical absorbing power of the medium at a particular wavelength at the same instant. The two techniques monitor the concentrations of different species.

plotting one concentration against the other: this representation gives the series of closed curves shown in the lower illustration. Which circuit the system describes is controlled by the rate at which the reactant A is supplied to the reactor.

Some actual reactions show periodic behavior that mirrors the looplike oscillation of a Lotka-Volterra system. However, in some cases the periodic behavior is considerably more complex. An example is shown on this page for a reaction involving bromate ions, chlorite ions (ClO_2^-), and iodide ions (I^-). Although the behavior is complex, in due course the same sequence of concentrations is retraced so long as the reagents are supplied.

THE BELOUSOV-ZHABOTINSKII REACTION

Now consider the original example of an oscillating reaction, the Belousov-Zhabotinskii reaction; this reaction has kindled much of the current interest in oscillatory phenomena in chemistry. The BZ reaction is essentially the oxidation of an organic compound by bromate ions (BrO_3^-) in an acidic solution. The reaction takes place in the presence of a catalyst that consists of a metal ion, and the key feature of this ion is that it should be able to occur in two states that differ by one electron: examples are the cerium ions Ce^{3+}/Ce^{4+} and the iron ions Fe^{2+}/Fe^{3+}. The oscillations of the BZ reaction are typically detected by adding a few drops of an "indicator," a substance that changes color according to the relative concentrations of the metal ions in their two states. Thus, we are essentially monitoring the reaction at second hand, by watching the indicator that monitors the pulsation of the metal ion backward and forward between its oxidized and reduced forms.

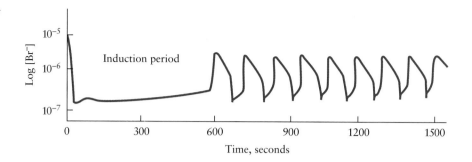

The periodic oscillation of the concentration of bromide ions during the BZ reaction. The concentration has been monitored electro-chemically using an electrode that is sensitive to the presence of Br⁻ ions. Note the lengthy induction period (of about 10 minutes) before the oscillations begin. The initial period of the oscillations is about 100 seconds, but it has decreased slightly on the right of the diagram because some of the BrO_3^- ions and malonic acid have been used up.

As we shall see, a crucial part of the mechanism of the oscillation is that there are two different types of redox steps, one involving only electron transfer (and hence catalyzed by the metal ions), and the other involving oxygen atom transfer (essentially from the bromate ion) and in which the metal ion catalyst plays no part. The overall mechanism was elucidated in 1972 by Richard Field, Endre Körös, and Richard Noyes, and involves a dozen elementary steps and a similar number of chemical species. We shall first describe the awesomely complicated "true" mechanism (true, that is, in so far as it is currently accepted), and then isolate the essential features in a greatly simplified version that is open to numerical analysis. This model system is called the "oregonator" (since Noyes and his group are at the University of Oregon). The following description of the mechanism is based on an excellent account given by Field and F. W. Schneider.

The overall mechanism of the BZ reaction consists of two main processes—Process A and Process B—that are connected by a linking gateway, Process C. The key to the oscillations will turn out to be that Process A and

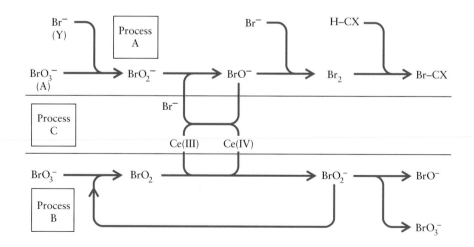

The network of reactions believed to contribute to the BZ reaction according to the Field-Körös-Noyes mechanism.

Process B alternate, and that by creating the conditions for each to take place, Process C passes control from one process to the other. The sequence of events is summarized in the illustration: although it may look overwhelmingly complex at first glance (and perhaps, very appropriately, more like an electrical circuit than a chemical reaction), it can be picked apart into manageable units, as we now show.

Process A can be regarded as a series of oxygen atom transfers from the bromate ion, BrO_3^-, to two bromide ions, Br^-, which culminates in the formation of molecular bromine, Br_2, the agent that attacks the organic compound. Each step of the process is enabled by the weakening of the bonds resulting from the presence of hydrogen ions in the strongly acidic solution that is used for the reaction. The first and second steps of Process A are both O-atom transfers:

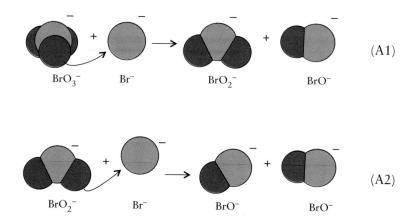

$$\text{BrO}_3^- \qquad \text{Br}^- \qquad\qquad \text{BrO}_2^- \qquad \text{BrO}^- \qquad\qquad \text{(A1)}$$

$$\text{BrO}_2^- \qquad \text{Br}^- \qquad \text{BrO}^- \qquad \text{BrO}^- \qquad\qquad \text{(A2)}$$

(In fact, since all the reactions take place in a strongly acidic medium, the protons that are present combine with the BrO^- and BrO_2^- ions to form the acids $HBrO$ and $HBrO_2$; however, for our purposes that is an unessential detail, and we need not show the protons in the mechanistic scheme.) Later we shall simplify these reactions to

$$A + Y \longrightarrow X + P$$

$$X + Y \longrightarrow 2P$$

with A denoting BrO_3^-, Y denoting Br^-, X denoting BrO_2^-, and P denoting BrO^-.

In the third step of the actual reaction, more bromide ions are transformed to molecular bromine, and the bromine goes on to react with the organic compound. In the classical BZ, reaction the organic reactant is malonic acid, $CH_2(COOH)_2$, and one of the hydrogen atoms is replaced by a bromine atom, to give $BrCH(COOH)_2$. However, these late stages of Process B are not crucial to the oscillatory character of the reaction, and we can consider P as the end of the line of that part of the reaction.

Process B is a counterpoint in the background of Process A. It springs into life once the earlier process has come to a virtual stop after exhausting the supply of Br^- ions, since Process B differs from A in the ability to take place in the absence of these ions. The process has two features that distinguish it from Process A: it involves at least one *electron* transfer redox step (as distinct from the oxygen-atom transfer that occur throughout Process A), and it is autocatalytic.

The first step in Process B is the reaction between BrO_2^- (produced in step A1) and BrO_3^- ions with the formation of the *radical* species bromine dioxide, BrO_2:

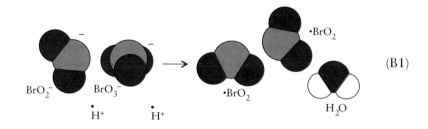

$$\tag{B1}$$

The metal ion now plays a role, passing an electron to a BrO_2 radical, reducing it to BrO_2^-, and in the process itself becoming oxidized. If the metal is cerium, this step in the reaction is

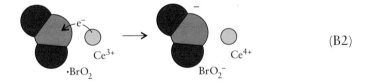

$$\tag{B2}$$

It follows that B2 takes the *two* BrO_2 radicals produced from *one* BrO_2^- ion in B1 and converts them into *two* BrO_2^- ions, so the net effect is the formation of two reactant molecules. We can combine the first two steps of the process into the overall reaction

$$BrO_2^- + BrO_3^- + 2H^+ + 2Ce^{3+} \longrightarrow 2BrO_2^- + 2Ce^{4+} + H_2O$$

The essential form of this reaction is

$$X + A \longrightarrow 2X + Z$$

where Z denotes the Ce^{4+} that is generated at the same time. The final step is the transfer of an oxygen atom between two BrO_2^- ions to form BrO^- and BrO_3^-:

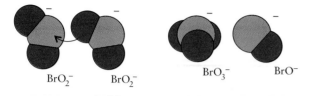

or, simply,

$$2X \longrightarrow A + P$$

Just as in Process A, the net effect of Process B is the reduction of BrO_3^- ions to BrO^-, although without needing to invoke Br^- ions. However, the side product of the reaction is the oxidation of cerium ions to Ce^{4+}.

The two processes are linked by Process C. To pass control back and forth between Processes A and B, we need a reaction that generates Br^- ions so that Process A can resume and also restores Ce^{3+} from its exhausted (that is, oxidized) form (Ce^{4+}) so that, when control is handed back to Process B, the metal ion can carry out the necessary reduction. It follows that the overall outcome of Process C must be *broadly* of the form

$$Ce^{4+} + BrO^- + \text{other Br compounds} \longrightarrow Ce^{3+} + fBr^- + \ldots$$

where f denotes the number of Br^- generated in the reaction. The abstract form of this reaction is

$$Z \longrightarrow fY$$

because Ce^{4+} ions (denoted Z) are removed and replaced by Br^- ions. We see that the reaction restores the cerium to its reduced form (Ce^{3+}) and leads to a substantial increase in the concentration of Br^- ions in the solution.

THE OSCILLATIONS OF THE OREGONATOR

It is easier to grasp why the oscillations of the BZ reaction occur if we examine the reaction in the oregonator. The key steps and their rate laws are collected as Processes A, B, and C on the next page.

Process A:	$A + Y \longrightarrow X + P$	Rate $= k_1[A][Y]$
	$X + Y \longrightarrow 2P$	Rate $= k_2[X][Y]$
Process B:	$X + A \longrightarrow 2X + Z$	Rate $= k_3[A][X]$
	$2X \longrightarrow A + P$	Rate $= k_4[X]^2$
Process C:	$Z \longrightarrow fY$	Rate $= k[Z]$

The origin of the oscillations can be understood by considering the role of Processes A and B separately, and then considering how Process C shunts control of the system between them. We shall suppose that the overall reaction is slow compared with the rate of oscillation sustained by the concentrations of the intermediates X and Y (BrO_2^- and Br^-). That means that the concentration of A (the BrO_3^- ion) is effectively constant on the time scale of the oscillations, and we do not have to consider the effect of the depletion of A.

At first, the solution may be rich in Y, and the overall reaction is dominated by Process A. However, this process consumes Y, and in due course there is insufficient Y for the process to play a significant role. Because A reacts more readily with Y (in Process A) than with X (in Process B), the latter process is barely active when the former is in progress. When Process A is dormant, Process B can take over. It leads (through its autocatalytic character) to a surge in the amounts of X and Z. With plenty of Z around, Process C becomes viable: it generates Y, which in due course rises to a sufficiently high concentration for Process A to reawaken and resume its dominant role. Then the cycle begins again and continues for as long as the reactant A is provided.

The oscillating concentrations of the intermediates X and Y can be plotted on a graph of Y versus X, just as in the discussion of the Lotka-Volterra system. As both concentrations change in time, they trace out a path that summarizes the behavior of the reaction. In this case, the arbitrary initial concentrations migrate on to a trajectory in the form of a closed loop. The trajectory shows that the concentrations return to the same values as they periodically increase and decrease. This loop, a "limit cycle," is an important new characteristic of this reaction: it shows that although the initial concentrations of the reactants may vary over a wide range, the reaction will still evolve until it displays the same periodic behavior. If a wide range of initial conditions evolves into the same limit cycle, the cycle is said to be "globally attracting." If only concentrations nearby the limit cycle evolve into it, then the cycle is only "locally attracting." After a perturbation, such as the sudden addition of bromide ions to the reaction vessel, the attracting trajectory will be jolted into a new configuration, but will return to its original periodic behavior of the limit cycle is globally attracting. The

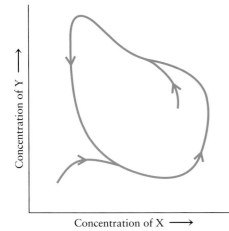

A limit cycle is a plot of the concentration of one substance against another (as in the looplike graphs shown earlier), and is the series of concentrations to which an arbitrary initial concentration migrates. The migration toward the limit cycle corresponds to the induction period shown in the illustration on p. 148, and the progress around the cycle corresponds to the oscillations.

limit cycle is an essential feature of any oscillating system, for otherwise the slight irregularities in concentration that are always present when a system is prepared would result in a state of unpredictable motion that would seem to be chaotic.

CHEMICAL CHAOS

One of the astounding developments of recent physical science—one that would not have been possible before the development of computers and their ability to pursue solutions of differential equations that are beyond the techniques of classical analysis—has been the discovery of systems of differential equations that have innately chaotic solutions. Thus, it is found that the solutions of such equations, although fully determined by the structure of the equations, are incapable of prediction. The kinetic equations that we have been considering are themselves of such richness that it should be hardly surprising that they—or minor elaborations of them—can display chaotic solutions. Instead of a reaction showing periodic oscillatory behavior, the concentrations burst into chaotic oscillation, and the concentrations of the intermediates show unpredictable amplitudes or unpredictable frequencies. Such behavior can be literally a matter of life and death, for should the heartbeat become chaotic, and the heart fibrillate, it may result in death. On a large scale, whole economies (which may also be described by differential equations for the flow of goods and money) may collapse into revolutionary chaos.

The successive period doublings that may lead to the onset of chemical chaos: after many such doublings, the limit cycle acquires a form that corresponds to the system showing unpredictable behavior.

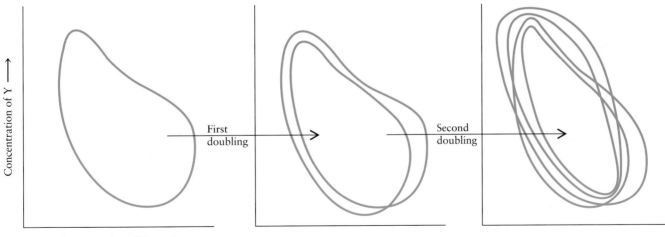

First doubling

Second doubling

Concentration of Y ⟶

Concentration of X ⟶

The onset of chaos can be displayed in a different manner by plotting the concentration of a single species at a series of times (the horizontal axis) against the concentration at a specific interval (in this case, 53 seconds) after each of those times. If the reaction is periodic, a simple curve would be obtained, but when the motion is chaotic, the curve is very complex. This illustration is for the concentration of Br^- ions in the BZ reaction after it has been steered into a chaotic regime.

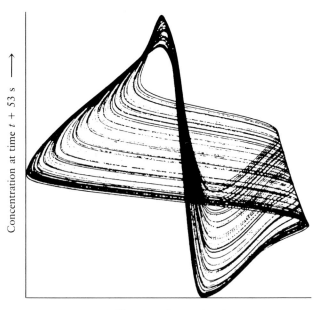

Concentration at time $t + 53$ s →

Concentration at time t →

The turbulent column of smoke rising from the candle flame is a chaotic system that is described by a strange attractor.

There are several ways in which a reaction can be steered into a chaotic regime and hence model a heart attack or an economic revolution in the safer environment of a test tube. For example, it may be the case in certain systems that as a parameter (such as a flow rate through the reaction vessel or the rate of stirring) is changed, the period of the limit cycle successively doubles, as shown in the illustration, and the system must circulate twice round the cycle before an initial pair of concentrations is restored. Then, after a further modification of the parameter, the period doubles again, and the limit cycle circulates four times before repeating itself. As each period doubles, the periodicity of the motion becomes less apparent, and finally appears like random fluctuations.

After this lengthy excursion into reactions in solution, we return briefly to the flame that started this line of thought, for unwittingly in his lectures, Faraday was demonstrating what only now can we understand: a flickering flame and its coiling column of smoke can display the onset of chaos. The differential equations that govern the flame and the rise of the smoke are complex, and involve hydrodynamics as well as chemical kinetics, but their underlying structure is closely akin to those we have been describing, and the transition from smooth combustion to turbulence is a sign that, lurking within them, is a "strange" attractor, a trajectory like that in the figure on this page where the behavior of the reaction is almost totally unpredictable.

PATTERNS AND PELTS

Finally, we take one further step into complexity, and consider the *spatial* structure that may emerge in unstirred chemical reactors. Here we come to the verge of understanding the reactions that give rise to the patterns so characteristic of animal pelts and butterfly wings.

Spatially periodic patterns can be modeled by taking into account the migration by diffusion of the reacting species. Thus, if the reaction vessel is not thoroughly mixed, a species may be produced in abundance in one region, and then spread by diffusion into a region where it can be eliminated by reaction with species that are abundant there. The spread of material from region to region may be used to explain the gracious spirals and circles produced by the BZ reaction in an unstirred, and therefore spatially inhomogeneous, reaction vessel. The concentration of Br^- ions, for example, is now inhomogeneous, since its concentration changes both as a result of Process A and as a result of diffusion through space. Thus, although control of the BZ reaction may be handed to Process B in a particular region, in another part of the system, the concentration of Br^- ions may still be high enough for Process A to be in command. The concentration of the Ce^{4+} ions produced in Process C will also be inhomogenous. It may readily be imagined that a complex *spatial* pattern of concentrations (and perhaps colors) will emerge from the reaction.

The complex patterns on the pelts of animals are the outcome of reactions that are periodic in space.

The spatial inhomogeneity of some reactions accounts for the appearance of patterns on animal pelts. The pattern on the integument of mammals develops toward the end of embryogenesis, but it probably arises from a pattern that is laid down considerably earlier. The canvas on which the pattern will later be laid down is formed by cells called melanoblasts that migrate over the embryo and in due course become melanocytes, the pigment-producing cells in the epidermis. The melanocytes generate the complex, light-absorbing molecule melanin, which gives a dark color, and the melanin then migrates from its site of synthesis in the hair follicle into the hair. Whether or not the melanocyte produces melanin depends on the presence of an as yet unidentified compound, and it is the diffusion of this compound that creates the pattern on the coat of the animal. Hence, the pattern is essentially an indicator that maps an underlying spatially periodic distribution of a metabolic product (the unidentified compound). The actual appearance of the pattern depends critically on the shape of the embryo, for the diffusion characteristics of the metabolic product are very sensitive to the geometric characteristics of the region in which it migrates.

Some calculated patterns for tapering cylinders (to represent tails) are shown in the illustration together with some actual tail markings. The appearance of the markings is a consequence of the shape of the tail in the embryo when the switching compound diffused and of the stretching that occurs as the embryo develops into the fully grown animal. For example, the markings on a leopard's tail are consistent with a reacting-diffusing system

Computed patterns of a spatially periodic reaction with diffusion over the surface of a tapering cylinder (top) and some drawings of actual animal tails (bottom). From left to right adult cheetah (Acinonyx jubatis), *adult jaguar* (Panthera onca), *pre-natal male genet* (Genetta genetta), *adult leopard* (Panthera pardus). *The small cone next to the leopard's tail is the form of the tail in the prenatal leopard.*

on the surface of a sharply tapering cone (the shape of the tail in the leopard embryo) that is subsequently stretched out into a nearly uniform cylinder. The spots go almost to the tip of the tail because in the sharply tapering conical surface stripes will only occur very close to the apex.

There are many reactions that show spatial periodicity—the emergence of form where previously there was none. The interesting general feature of these spatial patterns is that they correspond to the emergence of structure as a result (as we saw in Chapter 3) of collapse into chaos. Just as a waterfall gives rise to eddies and transient patterns, so chemical reactions that are far from equilibrium also give rise to patterns. We can therefore leave this apotheosis of the work that began a century and more ago in the collaboration of Essen and Harcourt with an appreciation of the intricacies of the gearing of the global collapse into chaos. Through its nonlinearity and feedback loops, the collapse generates such complexity that it is easy to be misled into considering that complexity designed.

THE VOYAGE OF TRANSFORMATION | 7

The rich tapestry of events in the living world, such as the germination of a seed, are the outcome of strings of individual events, as one atom is replaced by another, ejected from a molecule, or relocated within a molecule.

And now I have to ask your attention to the means by which we are enabled to ascertain what happens in any part of the flame; why it happens; what it does in happening; and where, after all, the whole candle goes to.

MICHAEL FARADAY, LECTURE 2

A candle that appears to be burning gently is, at the atomic level, the seat of violent eruption as molecules are ripped apart and new liaisons are formed. In this chapter we focus more closely on the enormous changes that occur as one molecule changes into another, and in doing so we come to understand the details of these events. We shall continue to follow the object of Faraday's gaze, but we shall see what modern knowledge would have added to his understanding of the flame, the origin and reactions of the fuel, and the subsequent chemical history of carbon dioxide and the compounds it goes on to form. There has come about a revolution in our understanding since Faraday's time: we no longer think merely of there being first one substance and then later another; now we can visualize the events that accompany the transformations of individual molecules—such as the departure of one atom and its replacement by another. It is through a knowledge of such details that we cease thinking of molecules as static entities, mere structures, and see them as dynamic, living, changing species that express the chemical personalities of the atoms from which they are built.

The reactions in the candle flame are abundant and complex, and even more so are the reactions of the carbon atoms liberated from the fuel, particularly if they are taken into the biosphere by photosynthesis: life is chemical complexity virtually without end. However, it is possible to cut through the maze of changes and identify three fundamental events: the *elimination* of an atom from a molecule, the *addition* of an atom to a molecule, and the *rearrangement* of the atoms of a molecule without net gain or loss. All three processes can also occur with entire groups of atoms, so a —Cl atom may be eliminated from a molecule, an $—NO_2$ group may be attached to a molecule, and an —OH group might migrate to a new location within a molecule.

As in a symphony composed of individual notes, these basic processes may occur consecutively (and sometimes simultaneously). The sequence in

CHAPTER SEVEN

which the processes occur is called the *mechanism* of the reaction. Suppose, for example, that in a flame or a biological cell, an incoming group of atoms attacks a molecule and causes a group already present to be ejected and then forms a bond in its place. The kinds of questions we need to answer before we can claim to understand a reaction fully include whether, when the molecule is under attack, it forms a bond to the incoming group before the elimination of the group to be replaced, at the same time as elimination, or only after elimination is complete. If rearrangement of the atoms occurs, when in the course of the reaction does it take place?

A tiny two-note chemical symphony, with one note elimination and the other note addition, is a *substitution* reaction. In this event, an atom (or group of atoms) is removed from a molecule, and another takes its place. One of the most useful substitution reactions is the replacement of an atom in a molecule by a carbon atom (together with any atoms that are attached to it), for in that way the carbon skeleton of the molecule can be extended and an intricate framework gradually built up. It is in this way that the chains of carbon atoms in the candlewax were grown originally in their living hosts as atom was linked successively to atom, and it is in this way that a carbon atom in carbon dioxide may, one day, resume its organic role after its reentry into the biosphere.

An idea of the intricacy of the changes that can occur in even a simple reaction is illustrated by the contortions that a hydrocarbon molecule goes through when it reacts with ozone, O_3. This reaction snips through a carbon–carbon double bond and slices a molecule in two. The reaction occurs in the atmosphere, where fragments of unburned hydrocarbon fuels are cut into smaller pieces by any ozone present. However, the double bond is not cleaved by a single hammer blow of a colliding ozone molecule, but by a delicate series of tiny shifts of electrons and nuclei. Indeed, the ozone reaction is typical of many chemical reactions in that it achieves change by

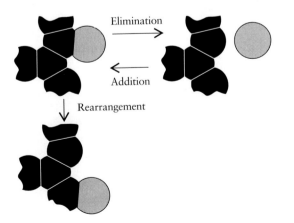

The three basic reactive events we consider are elimination, the removal of an atom from a molecule; addition, the attachment of an atom to a molecule; and rearrangement, the exchange of atoms initially present in a molecule.

minute steps. Very little happens at each tick of the reaction clock, but the overall effect may be a revolutionary change of structure.

The first step of the reaction is believed to be the joining together of the two molecules to form an intermediate. The ozone molecule retains its chainlike character, but its two outermost atoms link to the two carbon atoms initially joined by a double bond:

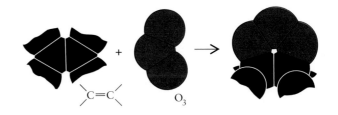

This five-membered ring structure is unstable, and bursts apart as the oxygen atoms suck electrons away from the stressed C—C bond:

Next, one of the severed halves rotates relative to the other:

Now only a small rearrangement of electrons is needed to form an intermediate that has actually been isolated, an "ozonide":

At this stage, the double bond has been replaced by two oxygen bridges and the carbon atoms are held well apart. The ozonide is ripe for fragmentation. A single collision in the atmosphere may break the bridges and cut the molecule in two.

CATEGORIES OF ATTACK

Although a carbon atom is subject to only three basic processes, these processes can take place in a variety of different ways. It is the exploitation of this mechanistic richness that gives the flame its power and carbon its structural variety and reactive fecundity. The mechanistic richness of carbon is also the root of the power that chemists have acquired to manipulate matter, for to conjure with carbon and its compounds they weave together a web of change from these simple mechanistic elements.

We can begin to see the potential for elaboration when we realize that even the simple act of breaking a bond to carbon may occur in three different ways. For example, each atom may retain one electron from the pair that formed the bond:

$$C-X \longrightarrow C\cdot \ + \ \cdot X$$

This process, which leads to radicals, is common in a flame. Alternatively, one of the two atoms may capture both electrons, to give either a positively charged carbon species, a *carbocation*,

$$C-X \longrightarrow C^+ + \ :X^-$$

or a negatively charged carbon species, a *carbanion*,

$$C-X \longrightarrow C:^- + X^+$$

Since the three different types of carbon species that may be formed will participate in distinctly different reactions, a particular outcome can be achieved by encouraging the bond to break in an appropriate way. It follows that we need to understand how each different type of bond breaking is encouraged or discouraged, for then we shall be able to understand the unfolding of the symphony of change in which carbon can participate. We shall see that both the intrinsic character of the molecule and its environment decide the outcome of a reaction: both nature (molecular structure) and nurture (the reaction conditions) determine the course of reactions.

To understand fully the personality of carbon, we need to understand the structural and environmental influences that can change the sequence of events in the reactions in which it is a participant and hence give rise to different products. In some cases the atom may be subject to attack by an incoming group. In others attack might be withheld until after the departure of another group has transformed the original molecule into an ion. That attack might even be withheld until after the ion has undergone a rearrangement. It is this deployment of the sequence of stages in a reaction, this construction of the score of the symphony from a handful of notes, that endows nature with its richness. It is through the careful sequencing of processes that a leaf or our body acquires its precise control over the construction of new molecules out of old.

For example, when can we expect reaction by radical formation, and when can we expect carbocations or carbanions to be involved instead? We saw in Chapter 3 that a polar solvent encourages dissolution of an ionic solid by emulating the environment of the ions in the crystal and effectively disguising the fact that the ions have separated. The formation of carbocations and carbanions resembles dissolution of an ionic compound, except that instead of an ionic compound being replaced by separated ions, a covalent bond is replaced by separated ions. As in true dissolving, the energy demands are minimized if the solvent can emulate an ionic medium, which it can (at least partially) if it is polar. Since the simulation cannot occur in a gas—where there is no solvent—ion formation in that state remains a very high energy route, and only rarely occurs. In the violently energetic flame of intense combustion, as in the heart of conflagration where the temperature is very high, there is energy enough to produce ions as well as radicals. However, most of the reactions in the gaseous conditions of a normal flame involve radicals.

The incipient formation of ionic species is common once the carbon atom has become a part of an organic molecule in the aqueous environment of a living cell. To a certain extent, water can readily emulate other ions, and by its pronounced ability to hydrate charged species, it encourages the formation of ions. It is on this rich reaction environment that we shall concentrate in this chapter, for this is where carbon displays its virtuosity most clearly. It must be understood from the outset that the reactions that carbon compounds undergo in living cells, as in photosynthesis or in the metabolism that leads to natural products such as the beeswax and spermaceti that Faraday used, are very elaborate and it would make a far too technical story to elucidate even one in detail. Instead, the following pages will show some of the individual steps that may occur in living cells and in laboratory emulations of certain types of reaction. We shall consider some of the building blocks of change, the individual notes (and sometimes a single bar) of the symphony of change, not the entire concert.

ELECTROPHILES AND NUCLEOPHILES

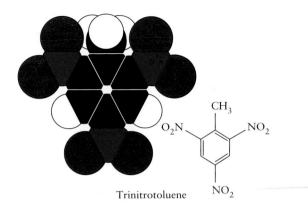

Trinitrotoluene

The outcome of a reaction often stems from the location selected by the incoming group for its attack on the molecule. For example, that group may be an *electrophile*, a group that, perhaps because of its positive charge, seeks out centers of electron richness. A double bond in an organic molecule represents an electrically juicy region of the molecule that an electrophile might select to attack. On the other hand, if a region of a molecule is particularly thin in electrons, so that the positive charge of a nucleus shines through like the sun through a mist, then that may be the site which a *nucleophile*, a nucleus-seeking species, will select as a target. A ripple of electrons at one location may ensure that that is where an electrophile will strike. If the veil of electrons is drawn back from another region, the revealed nuclear charge may act like bait to attract an incoming nucleophile.

Electrophiles are often positively charged species, for positive charge seeks out regions of high electron density. An example is the NO_2^+ ion, which chemists sometimes use when they want to make a C—N bond, as in the manufacture of the high explosive trinitrotoluene (TNT) from toluene: the positive charge of the NO_2^+ ion seeks out regions of high electron density in the toluene molecule. However, nature is subtle, and an electrophile is not necessarily a positively charged species. Thus, electrophiles also include the uncharged bromine molecule, Br_2. The outermost of the 35 electrons of a bromine atom are gripped quite feebly by the nucleus buried deeply under its closed shells, and hence the outermost electrons in the bromine molecule are moderately mobile and can respond to the presence of a high electron density. Thus, the region of high electron density on the molecule that is being attacked forges the weapon that is used to attack it.

The electrophile par excellence is the proton: being so small, it can wriggle into the tiniest crevices of molecules and attach to almost any atom (and especially to oxygen and nitrogen atoms, with their available lone pairs). The proton plays a very special role because it can attach to almost any molecule, and by bringing about an adjustment of the molecule's electron distribution, sensitize it to further attack or initiate a rearrangement of its atoms. The proton is the great *enabler* of reactions, the key that opens the door to change. That is one of the reasons why the maintenance of the acidity of our bodily fluids—particularly of our blood—is so essential to health, for the reactions that constitute life depend critically on the availability or otherwise of protons, and their absence or undue abundance may inspire the series of reactions we construe as disease and death.

Just as an electrophile is not necessarily a positively charged species, a nucleophile is not necessarily a negatively charged species. However, a negative charge does add an edge to a nucleophile's attack, and some of the best

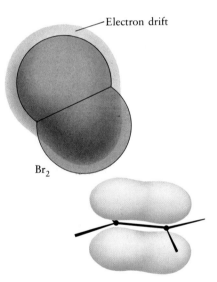

Electron drift

Br_2

As a bromine molecule approaches a region of a molecule where the electron density is high, its electrons are pushed back from the front of the molecule (because like charges repel), so revealing the positive charge of the bromine nucleus within. The exposed positive charge of this nucleus is attracted toward the electron-dense region of the double bond.

nucleophiles include negatively charged species such as the hydroxide ion, OH^-. This ion forms C—O bonds when it attacks, as when carbon dioxide dissolves in water and carbonic acid is formed.

ELIMINATION REACTIONS

Even though we may be able to dissect a reaction into three elementary processes, each one of those processes is itself a composite and possibly complex sequence of events. Consider, for instance, an ostensibly simple event, the elimination of a pair of atoms from neighboring carbon atoms, and their replacement by a carbon–carbon double bond:

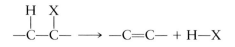

This reaction is called a "1,2-elimination," the consecutive numbers indicating that atoms have been ejected from adjacent sites.

There may be several reasons why it may be appropriate for an enzyme (or a chemist) to introduce a double bond into a molecule. A double bond introduces a center of enhanced electron density that can act as the site of electrophilic attack in a subsequent reaction. It also brings a structural rigidity to a molecule that is absent when the atoms are joined by single bonds, for although one part of a molecule can rotate relative to another part to which it is attached by a single bond, that is not the case when the two parts are joined by a double bond, for a double bond confers torsional rigidity on a molecule. Torsionally rigid molecules typically have physical properties that differ from those of floppy molecules, and may also undergo different reactions. One of the physical differences is that large molecules with double bonds are structurally less pliable than those with single bonds, and pack together less well; hence long-chain esters with several double bonds are oils, whereas their analogs with fewer double bonds, or only single bonds, are the solid fats and waxes from which Faraday's candles were made.

We shall explore the mechanism of elimination reactions by considering two extreme possibilities. First, we consider an "E1 reaction." In such a reaction, there is only one molecule (hence the 1) involved in the rate-determining step of the reaction. To imagine what goes on during an E1 reaction, we should think of the first step as the molecule falling apart to form a carbocation:

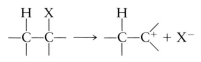

The edible-oil industry can control the consistency of its products by controlling the number of double bonds in the substances it prepares. When the double bonds in the hydrocarbon tails of the molecules that make up the oil (top) are replaced by single bonds, the molecules pack together better and result in a solid fat.

This ionization is usually slow, because a considerable energy investment is required before it is feasible. As we remarked earlier, the energy investment can be minimized and the step encouraged by using a solvent that is strongly polar.

The structure of the molecule itself also plays a role in determining whether it is likely to form a carbocation. The net positive charge of a carbocation notionally resides on the carbon atom that has lost the group, and so that atom is a center of attraction for the electrons that remain in the molecule—it is, in a sense, a center of electrostatic stress. This stress can be partly relieved if electrons in the vicinity of the C^+ center can migrate toward the positive charge. It follows that the presence of a nearby group that can release electrons at the demand of the C^+ center will encourage the formation of a carbocation. Hydrocarbon groups are of this kind, for they are reasonably rich in electrons and do not have electronegative atoms that would compete for the electrons.

Once the carbocation has formed, and the departing group has diffused some distance away, the second reactant—which is typically a Lewis base— can move in to round off the attack. Its electron pair is attracted toward the hydrogen atom on the carbon atom next to the center of positive charge. This is because the positive charge sucks electrons away from the hydrogen atom, partly denuding it of its share in the bonding electron pair, and so leaves the proton charge to show through the pale mist of electrons and render it open to attack. Moreover, because the electron pair has drifted away from the hydrogen atom, the carbon–hydrogen bond has been weakened. The hydrogen atom is ripe for harvesting. (The attacking group also approaches the positively charged carbon atom, and hence gives rise to a competing substitution reaction of a kind we describe later: chemical reactions, particularly the complicated reactions of organic chemistry, rarely result in the formation of a single product in 100 percent yield.)

The Lewis base :B homes in, like a guided missile, on the partially positively charged hydrogen atom, spears it with its electron pair, and carries it off as H—B, leaving the electron pair that originally held the hydrogen atom in the sole possession of the carbocation. The hydrogen atom, which has departed as a proton, has carried off a single positive charge, with the result that the carbocation becomes a neutral species once again. The deserted lone pair is now free to shift into the region between the two carbon atoms, and hence to form the second component of a double bond.

The series of events that we have just described occur essentially simultaneously. Once the carbocation has formed—that is the uphill part of the reaction, and the step that controls its overall rate—the Lewis base can approach very rapidly under the homing guidance of the carbocation's positive charge. At the same time as the proton departs, the electron pair that held it to the carbon atom slips into position and takes up its new task of

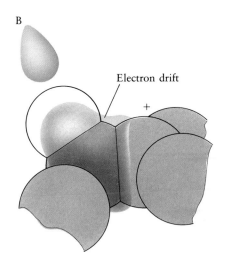

The electron pair of a Lewis base is attracted toward the hydrogen atom on the carbon atom because the nuclear charge of the hydrogen has been exposed. Since the C—H bond has been weakened, the incoming base can carry off the proton.

helping to bind the two carbon atoms together as one component of a double bond:

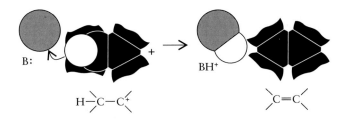

Next, we consider how an elimination reaction may occur when a species is very reluctant to ionize, either because there is no tappable reservoir of electrons in the molecule—so that the carbocation is intrinsically not very stable—or because the reaction medium is inhospitable to ionic species. An elimination reaction may still take place, but its mechanism is closer to being that of an "E2 reaction," in which *two* molecules are simultaneously involved in the formation of a fleetingly transient cluster of atoms. Now the departure of one group and the spearing of the proton off the neighboring carbon atom occur at the same time, and no discrete carbocation is formed at any stage.

To visualize the formation of the products from the reactants we should think of a single, gliding progress through the reaction, with electrons adjusting to the extraction of the proton and the expulsion of the other group. We could picture the lone pair of the attacking Lewis base as impelling electrons before it through the molecule, like a ram, until they make their escape by riding off as an X^- ion:

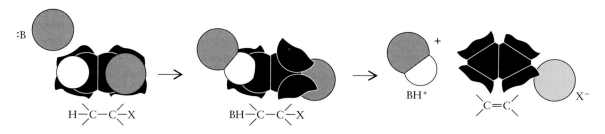

THE STEREOCHEMICAL CONSEQUENCES OF ELIMINATIONS

Both the E1 and E2 mechanisms are routes from products to reactants, as are the mechanisms that lie between these extremes, with marginally different choreographies. At first glance the products would be expected to be the

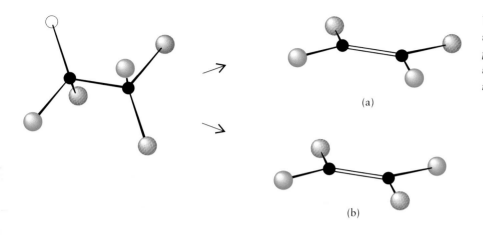

(a)

(b)

same by either route. However, chemistry is not a subject of first glances, and in fact there may be an important difference between the products of reactions that have occurred by different routes. In particular, the products might differ in their stereochemistry—the arrangement of their atoms in space.

For instance, suppose we started with the molecule depicted in the illustration: there are two possible products, (a) and (b), and one cannot be rotated into the other because the double bond confers rigidity. One of the products is obtained if the new double bond forms when the two parts of the molecule are in one relative orientation and are then locked into that orientation by the newly formed, torsionally rigid bond. The other product is obtained if the double bond is formed when the two parts of the molecule are in the alternative relative orientation. Such stereochemical specificity can have the profoundest consequences: one of the most tragic examples is thalidomide, for which there are two versions that are the mirror image of each other. The commercial product was a mixture of both forms, and one of those forms induced congenital deformations; had the production led to the other form alone, widespread personal tragedy would have been avoided.

In an E1 reaction, the carbocation is formed before the double bond is able to develop, for the proton on the neighboring carbon atom acts like a pin holding back the relocation of the electrons, and the C^+ end of the molecule is free to rotate around the single bond. When the "pin" is removed, the double bond springs into existence and the current orientation of the C^+ ion is trapped. Since the double bond may be formed at any moment, a mixture of the two products will be formed with similar abundances.

So long as the proton is present to hold back the flood of electrons into the bonding region, the other part of the molecule can rotate freely around the bond. However, as soon as the proton leaves, the electrons can flow into the bonding region to form the second component of a double bond, and the molecule is trapped into its current orientation.

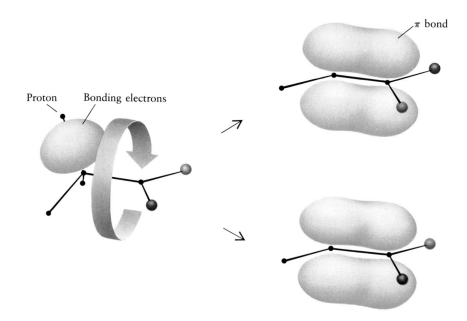

Proton   Bonding electrons   π bond

The outcome of an E2 reaction is quite different, for the intermediate cluster of reactants that corresponds to one orientation may be significantly easier to form than that for the other orientation. In fact, it generally turns out that the departing group X⁻ leaves from the *underside* of the complex, the side away from attack by the Lewis base. That is, the hydrogen atom and the other group are removed in opposite directions, as shown in the illustration at the top of the facing page. The implication of this mechanism is that there will be only one of the two possible products formed, specifically product (a).

ADDITION

Now we turn to the opposite of elimination, addition to a double bond:

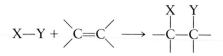

A double bond is the scar left after a pair of neighboring atoms have been ripped from a molecule in a 1,2-elimination reaction. Since that scar is an electron-rich region, it may function as a target for an electrophilic addition

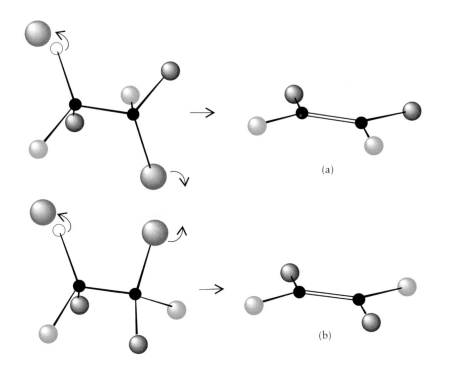

(a)

(b)

reaction in which the double bond is ruptured and a pair of atoms is attached in its place. It should be easy to appreciate that a judicious sequence of elimination and addition reactions may achieve the *overall* reaction:

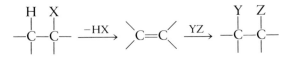

and so two reactions performed in sequence can result in a substantial change to a molecule.

The interesting feature of an addition reaction is the influence that the medium can exert on the stereochemistry of the product. As an illustration, we consider a reaction in which bromine is added to neighboring carbon atoms:

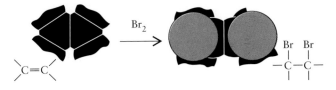

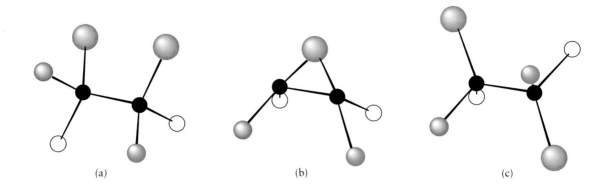

(a) (b) (c)

If a Br₂ molecule were to add to a double bond directly, it would result in the formation of a molecule with the structure shown in (a). However, since the formation of the intermediate (b) protects one side of the molecule from attack, the product of the reaction is in fact the molecule (c). Although one group of the atoms can rotate relative to the other around the C—C bond, rotation of (c) does not result in the formation of (a).

The addition of bromine atoms to an organic molecule is a way of sensitizing the molecule to further attack, and hence to prepare it to be a reactant in a subsequent step in an overall reaction scheme. The reaction is unimportant in living systems, because bromine molecules do not occur naturally, but it is used in chemical industry as a stepping stone to the construction of elaborate molecules. The striking feature of the reaction is that the addition of bromine does not give the product (a) that would result if the two atoms approached from the same side of the double bond: the stereochemistry of the product is explicable only if the atoms approach the molecule from opposite sides of the molecular plane.

The mechanism that is consistent with this observation is that an incoming Br₂ molecule acts as an electrophile and attacks the region of high electron density at the double bond. As the electrons in the bromine molecule drift back from the leading bromine atom, the Br—Br bond lengthens and the trailing bromine atom breaks off, taking the bonding pair of electrons with it. The resulting organic molecule has captured the leading bromine atom (as a Br⁺ ion), and the species (b) has formed. The bridging bromine atom protects its side of the cation against attack by an incoming Br⁻ ion; when attack finally occurs, the bromine atom attaches to the opposite side of the plane to the bromine atom already present, so forming species (c) almost exclusively.

SUBSTITUTION

In a substitution reaction, one atom in a molecule is replaced by another:

$$C{-}X + Y \longrightarrow X + C{-}Y$$

The red-brown liquid element bromine is produced by the oxidation of an aqueous solution of bromide ions, Br^-, by chlorine gas. It is used in industrial chemistry in a wide variety of synthetic procedures because the presence of a bromine atom in an organic molecule sensitizes it to further attack.

Substitution reactions are common procedures for the formation of a variety of carbon-element bonds and hence for building up increasingly sophisticated molecular structures. For example, a CN^- ion may participate in the substitution reaction

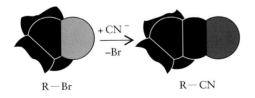

and hence create another C—C bond.

We shall confine our attention to nucleophilic substitutions in this part of the discussion, but move on to other types (particularly electrophilic substitutions) later. All the reaction mechanisms may be designated S_N, where S stands for substitution and N for nucleophilic. When mechanistic organic chemistry emerged in the first half of this century, two principal modes of nucleophilic substitution were recognized: unimolecular substitution and bimolecular substitution. They have since been modified, but they will serve as an introduction to the visualization of the events that accompany reactions like these. As we shall see, just as we did in the discussion of elimination and addition reactions, each class has particular stereochemical consequences, so knowing when to avoid one route and how to encourage another may lead to products with different properties.

One mode of attack to imagine is designated S_N1 (and now more circumspectly called "limiting S_N1"). It begins in the same way as the analogous E1 elimination—indeed, that the reactions begin in the same way accounts for the fact that eliminations often complicate substitutions by occurring with them. The initial slow step is the departure of a group to leave behind a carbocation:

$$R—X \longrightarrow R^+ + X^-$$

Once the carbocation has formed and the leaving group has left its vicinity, the nucleophile can move in to round off the attack. It may approach from either side of the plane of the molecule, and hence give rise to either of the two products below labeled (a) or (b) in approximately equal abundance.

The second mode of attack we should try to imagine is quite different. In it, there is a *concerted* motion of the entering and leaving groups, and the new bond is forming as the old bond is breaking. This mechanism is denoted S_N2, because (like the analogous E2 reaction) two molecules—the initial molecule and entering nucleophile—are involved in the rate-determining stage.

A major difference between an S_N1 and an S_N2 reaction is that in the latter an exquisite level of control is exercised over the stereochemical out-

The carbocation intermediate in an S_N1 reaction is planar, and can be attacked from either side of the plane. As a result, the products (a) and (b) will be formed in almost equal abundance when the gray group attaches to the intermediate.

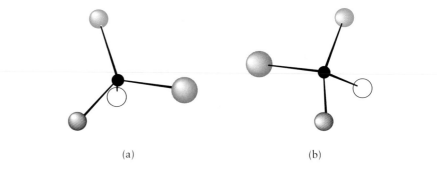

(a)   (b)

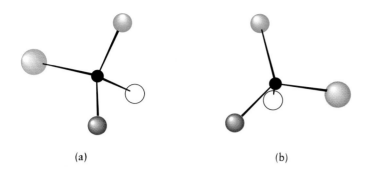

(a)   (b)

In contrast to the outcome illustrated on the opposite page, there is only one product of an S_N2 reaction. If the starting material is (a), then the incoming group attacks from the diametrically opposite side of the molecule, and the product is (b): the umbrella-like molecule undergoes inversion.

come of the reaction, for as the entering group enters and the leaving group leaves, the molecule effectively inverts like an umbrella turned inside out by the wind. Therefore, if the starting molecule, labeled (a) at the top of the page, the outcome will be (b) and not its umbrella-inverted partner. That is, whereas an S_N1 reaction gives a *mixture* of products, an S_N2 reaction gives a *single* product with the inverted stereochemistry.

THE QUANTUM MECHANICS OF ATTACK

One very interesting question concerning an S_N2 reaction, which like so much in chemistry does not have a simple answer, is why an entering nucleophile attacks from behind the molecule (on the side opposite to the C—X bond), and does not sever the old C—X bond by a frontal attack. The explanation appears to depend on a purely quantum mechanical effect. Such effects will be of central importance in the next chapter, and the rest of this chapter will act as bridge to that material.

To understand what is happening we must return to the discussion of wavefunctions, orbitals, and bonding that was sketched in Chapter 2. There we saw that a bond is formed when an orbital on one atom overlaps an orbital of another atom, and there is an accumulation of electron density between them when the resulting molecular orbital is occupied by two electrons. Now consider what happens when a nucleophile—which we can represent as a σ orbital with a lone pair of electrons in it—approaches the C—X bond. The latter is actually a *full* σ orbital, so it cannot accept any more electrons.

However, we should not jump to the conclusion that the bond is protected because its orbital is full, for not far above the full σ orbital there lies an empty antibonding σ orbital that in principle can accommodate two more electrons and hence enable the molecule to act as a Lewis acid and

The approach of a nucleophile to a full bond leaves overlap with the empty antibonding partner as the only possibility; however, the internuclear node results in zero net overlap, and this approach is unsuccessful. An approach to the molecule from the side opposite to the bond can result in net overlap, and as a bond is established, the cluster of atoms becomes planar and finally inverts.

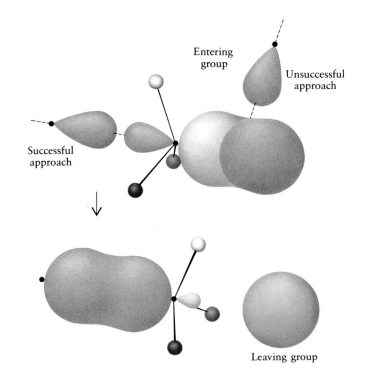

accept an electron pair. But now another trick up nature's sleeve comes into play, for there is also a symmetry requirement that must be satisfied. In the next chapter, symmetry will take center stage, but even here we can begin to see its crucial but often unrecognized significance.

Consider the shape of the incoming orbital occupied by the lone pair: it is a σ orbital around the line of attack. However, the empty antibonding orbital, although it has σ symmetry around the C—X direction, has very close to π symmetry around the line of attack. Therefore, whereas the incoming orbital has constructive overlap with one half of the orbital it needs to bond with, it has a destructive overlap with the other half. As a result, there is little net bonding between the incoming group and the molecule it is attacking: symmetry is a defense against attack.

The alternative attack, from behind, is symmetry allowed. When the orbital of the incoming group lines up with the molecule, it overlaps with a lobe of the antibonding orbital, having σ symmetry around the line of attack. It can thus form an incipient σ bond which progressively strengthens as the leaving group departs, the molecule inverts, and the antibonding orbital becomes more accessible. Here we have the first example of a reaction that is governed by symmetry and interference between wavefunctions. We shall see more of this deep structure of reaction mechanisms later.

ELECTROPHILIC SUBSTITUTION

Natural products that contain benzene rings occur widely in nature, and the mode of their construction is in many cases by an enzymatic reaction in which three acetate ions, $CH_3CO_2^-$, are pinned together to give a six-membered ring of carbon atoms. Thus, one route for carbon dioxide to become a component of a benzene ring is for the carbohydrate that it forms during photosynthesis to break down partially into acetic acid, and then to be incorporated into a ring.

Electrophilic substitution is the dominant mode of reaction for compounds containing electron-rich benzene rings. Therefore, when we turn to ways in which it is possible to decorate the benzene ring with other groups (as in the production of TNT mentioned earlier), we need to explore how an electron-seeking group attacks it. The classic reaction (in the sense that much of our initial understanding of the reactions of benzene came from its study) is the attachment of an —NO_2 group to the ring in place of a hydrogen atom, and we shall introduce the subject with this example. One question that arises is the position at which an entering group attacks the ring relative to any other group that is already attached to it. For example, if we were aiming to nitrate toluene, which already has one substituent in the ring, would the —NO_2 group attach at a neighboring position, one place further round the ring, or elsewhere on the ring? Indeed, how does nature aim a molecule-sized dart at a precise position on a molecule?

The nitration of benzene in the laboratory is carried out by warming benzene with a mixture of concentrated nitric and sulfuric acids. It is known that such a mixture contains a small proportion of the nitronium ion, NO_2^+, and that this species is the attacking agent. Because it has a positive charge, the nitronium ion seeks out a region where high electron density can accumulate—which it finds at any of the carbon atoms of the ring. As it attacks, it forms a C—N bond while weakening the C—H bond at the same location (see the illustration on the next page). The hydrogen atom departs as an ion, and the C—N bond settles down to its final strength.

The accelerator, brake, and steering wheel that nature has at its disposal for controlling the speed and direction of an incoming electrophile is the character of the substituent already present and, in particular, its ability to supply or withdraw electrons from the ring as the electrophile approaches. If the substituent is a hydrocarbon group like —CH_3, it is a source of electrons, just as we saw in our earlier discussion of carbocations. Therefore, the presence of groups such as —CH_3 and —$C(CH_3)_3$ act as accelerators of electrophilic attack. On the other hand, a substituent such as a halogen atom sucks electrons out of the ring, and hence acts as a brake on the reaction. The effects can be large; for example, the rate of nitration of

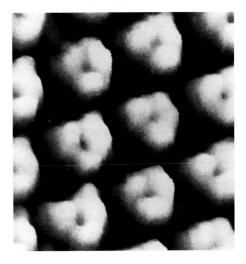

This is an image of benzene on the surface of graphite. The hexagonal shapes of the molecules can be made out quite clearly by means of scanning tunneling microscopy.

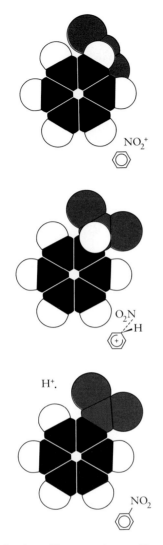

In the nitration of benzene, the attacking electrophile NO_2^+ approaches a carbon atom of the ring and begins to form a bond with it. At the same time, the hydrogen atom attached to the same carbon atom bends away from the ring and its bond is weakened. As the C—N bond becomes stronger, the C—H bond becomes still weaker, and at the end of the process the —NO_2 group is attached to the ring and the proton has been carried off by the solvent.

toluene is 30 times larger than that of benzene itself, and the rate of nitration of chlorobenzene is 30 times slower.

Controlling the overall rate is not very important, except in certain instances. Much more to the point is that it is possible to steer the reactant to a particular location, for it is possible to control the reaction rate at *selected* locations around the ring by attaching to the ring a substituent that can supply electrons in a definite way. Once again, this ability to control a reaction stems from quantum mechanics, but it can be expressed pictorially.

To see the full story, we need to step back in time a little, and indeed, step aboard a London omnibus. For there, so anecdote has it, did Friedrich Kekulé von Stradonitz have a dream that led him to propose, in 1858, the now famous hexagonal structures of benzene, shown in the margin. The actual structure of benzene is now recognized to be a blend of many structures, the two most important being these two "Kekulé structures." A good first approximation to its description is as a "resonance hybrid" of the two. In quantum mechanical terms, the structure of the benzene molecule is a sum of the wavefunctions for the two Kekulé structures, with smaller contributions from other structures. Thinking of the structure as a blend of the two Kekulé structures in equal proportions is adequate for our purposes.

Now consider the resonance structures of chlorobenzene. The two most important, in the sense that they make the dominant contributions to the overall description of the molecule, are the ordinary Kekulé structures but with the replacement of one hydrogen atom by a chlorine atom. There are also additional contributions from other structures that we can recognize once we note that a bound chlorine atom, —$\ddot{\underset{..}{C}l}$: , has three lone pairs of electrons. At least one of these pairs may move into a bonding location between the carbon and chlorine atoms and hence give rise to a double bond. The pair of electrons brings a negative charge into the ring (the electron pair is shared by the two atoms, so effectively *one* electron is transferred to the ring). As we see from the structure shown as (b), the resulting structure has a negative charge on the atom next to the carbon atom to which the chlorine atom is attached: this is the *ortho* position. In fact, there are two equivalent structures, since there are two *ortho* positions (b) and (c).

Yet another structure arises from the donation of the electron pair by the chlorine atom, and that one ends up with a negative charge on the carbon atom diametrically across the ring, the *para* position, as in (d). However, we cannot play with the bonding arrangement to achieve a negative charge at the remaining two (equivalent) positions, the *meta* positions (unless we make use of highly excited electronic states). This very important point means that the participation of a lone pair of a substituent atom in the bonding π system of the ring—which is called the "mesomeric effect"—accumulates electrons and hence negative charge) *specifically* at the *ortho* and *para* positions, not at the *meta* position.

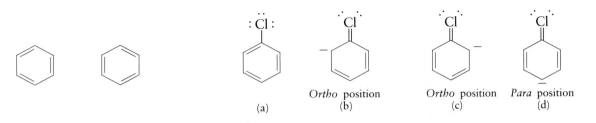

Kekulé structures of benzene

Ortho position
(b)

Ortho position
(c)

Para position
(d)

Kekulé structures of chlorobenzene

Since the mesomeric effect increases the electron density at the *ortho* and *para* positions, it encourages an electrophile to attack at those positions, and its ability to supply electrons to those locations as the incoming group attaches to the ring helps to stabilize the new bond. Therefore, when chlorobenzene is nitrated, we can expect to obtain principally the products having —NO_2 in the *ortho* and *para* positions, and hardly any molecules having —NO_2 in the *meta* position. That is what is observed in practice.

We have seen what influences the relative proportions of *ortho*, *meta*, and *para* products, and how a substituent already present can effectively eliminate the formation of *meta* relative to the other two. But can nature steer the dart toward the *ortho* position alone? That is not too difficult, once we realize that it may be possible to induce the incoming electrophile to glide past a substituent that is already present, and hence be encouraged to attack at a position close to the point of that substituent's attachment.

For example, suppose the substituent has an oxygen atom close to the point of attachment, as is shown in the illustration in the margin. The incoming nitrating agent may momentarily attach to the oxygen atom as it migrates toward its target, and then leap from its point of attachment to the nearest carbon atom in the ring. Thus, the reaction gives an almost pure *ortho* product: the unseeable dart has been directed to the ring at a position barely 100 pm from an alternative site. *That* is control over matter.

As we stressed earlier, we have sought to convey some insight into the individual steps that contribute to a few of the reactions that a carbon atom undergoes. When the atom escapes from the hydrocarbon that locks it into the candle, passes through the atmosphere as carbon dioxide, and then becomes a carbohydrate, it enters the milieu where it will be subject to the processes we have described (and, inevitably, many more). The chemical history of a carbon atom is a sequence of tiny changes, of atoms being driven out and other atoms forming new bonds. These steps occur under the rigorous grip of enzymes that conduct the symphony of changes, and they might, in due course, result in the carbon atom ending up as wax again. Then the taper of a later Faraday can send it on a new and infinitely varied voyage of transformation.

A group of atoms may act as a staging post for an attack by an incoming group and direct the latter to a specific location in a molecule. In this example, the electrophile attaches briefly to the oxygen atom (red) of the substituent before making its final approach to the benzene ring and attaching permanently to the neighboring carbon atom.

SYMMETRY AND DESTINY | 8

Symmetry, represented here by the orderly hexagonal structure of a honeycomb, is one of the factors that controls the outcome of chemical reactions. As we shall see in this chapter, symmetry permits some reactions but forbids others.

There is another little point which I must mention before we draw to a close—a point which concerns the whole of these operations, and most curious and beautiful it is to see it clustering upon and associated with the bodies that concern us . . ."

MICHAEL FARADAY, LECTURE 6

We have seen a little of the intrinsic quantum mechanical character of chemical reactions. The tip of this iceberg was visible in Chapter 7, where we saw that a combination of the Pauli principle and the characteristic shape of an antibonding orbital prevented frontal attack on a molecule and left the incoming group to approach from behind. Now we shall allow this concept to flourish as we see that there are certain reactions that are impossible to understand if we do not use quantum mechanical principles.

The classical principles of mechanics with which Faraday was familiar are almost totally silent on the mechanism by which chemical reactions take place. However, the reaction mechanisms that we explore in this chapter depend critically on the quantum mechanical characteristic that in Chapter 2 we called the "color" of an electron distribution (and by which, more formally, we mean the sign of its wavefunction).

In addition to quantum mechanics, there is another set of ideas needed to support the explanation of the reactions we shall meet; although these ideas were emerging during Faraday's lifetime, they were fully formed only well after his death. Group theory—the mathematical theory of symmetry—emerged from the highly abstract but epoch-making work of the youthful French mathematician Évariste Galois on the theory of equations, shortly before a duellist's bullet felled him in 1832 at the age of 21. Such abstract considerations as Galois developed would have had no direct impact on Faraday even though we now see that they are central to our comprehension of the nature of matter and, ironically but fittingly for Faraday, of the position of electromagnetism in the cosmic stable of forces.

Symmetry considerations had very little impact even on scientists with a greater taste for mathematics than Faraday, until the work of the German mathematician George Frobenius at the turn of the century, almost exactly a hundred years ago. Frobenius effectively transformed an abstract theory applicable to algebraic equations into a practical description of the symmetry of real objects. Nevertheless, despite the crucial transition from the decidedly abstract to the potentially practical, symmetry was not widely used to rationalize observations and detect relations between the apparently disparate until after the emergence of quantum mechanics in the 1920s, largely through the work of Eugene Wigner and Herman Weyl in Göttingen and Princeton.

The transfusion of group-theoretical thought into chemistry took longer. Among the first purely chemical applications of the quantitative concepts of group theory was its use in the description of the structures of molecules in the 1930s. The centrally important step for our purposes—the application of symmetry arguments to chemical reactions—did not firmly take place until the work of the American chemists Robert Woodward and Roald Hoffmann in the 1970s.

We shall continue the tradition begun in Chapter 7 of focusing on fragments of complete reactions, seeing how a particular atomic event can be enacted, and recognizing that an overall reaction may be the composite of many steps. To achieve a definite object a chemist—including nature's chemist, a biological cell—may carry out one of the reactions we are about to describe as but one step in a controlled avalanche of transformations.

One class of reactions that exhibit quantum mechanical effects most sharply puzzled chemists for many years, since they showed few of the responses that had become familiar in connection with reactions like those described in Chapter 7. On the one hand, their rates are largely unresponsive to changes in the polarity of the solvent; this suggests that carbocations

Robert Burns Woodward (1917–1979).   *Roald Hoffmann (b. 1937).*

and carbanions are not involved. On the other hand, their rates do not respond to species that help to generate radicals, which suggests that radicals are not involved either. Moreover, despite strenuous efforts, no reaction intermediates have been isolated, which suggests that none is formed and that the reactions take place in a single sweeping step.

One important example of these enigmatic reactions is the "Diels-Alder reaction." This reaction was identified in 1928 by the German chemist Otto Diels and his assistant Kurt Alder, who were awarded the Nobel Prize in 1950 for their discovery. In the Diels-Alder reaction, a ring of six carbon atoms is created by a reaction of the form

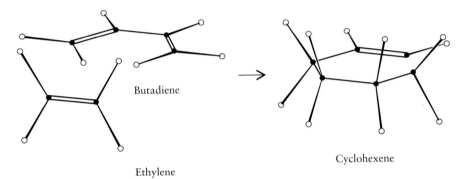

Butadiene

Ethylene

Cyclohexene

The reaction takes place very readily; it is also widely applicable in the sense that it occurs between many species that have the same pattern of bonds as shown here. Because of its facility and generality, the Diels-Alder reaction is of great utility for the synthesis of complex products—specifically for building rings of carbon atoms—and is used in the synthesis of a wide range of natural products, including steroids and vitamins.

The Diels-Alder reaction and the other reactions we shall consider are examples of "pericyclic reactions." The name conveys two aspects of their mechanism. One is that they take place by a "concerted process"—one in which bond breaking and bond formation occur simultaneously (this is consistent with the failure to detect any intermediates produced in the course of the reaction). The second is that they take place by the migration of electrons within a ring of interacting atomic orbitals. The latter point will be explained shortly, and will be central to this chapter.

We shall consider two types of pericyclic reaction. In a "cycloaddition reaction" two molecules come together to form a ring and form new σ bonds at the expense of breaking old π bonds. This is the basic pattern of the Diels-Alder reaction, and it is in the conversion of π to σ bonding that we shall find the quantum mechanical effects that determine the outcome of the reaction. The other type of pericyclic reaction, called an "electrocyclic reaction," is the opening or closure of a ring in a *single* molecule. An example is the ring-closing reaction in which one end of a butadiene molecule bends round on itself, forms a bond, and emerges from the reaction as cyclobutene:

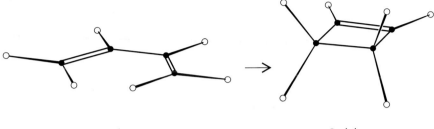

Butadiene   Cyclobutene

Because only one molecule is involved in this reaction, we shall consider electrocyclic reactions first.

ELECTROCYCLIC REACTIONS

The effects of quantum mechanical principles in pericyclic reactions are best revealed by exploring a particular question: What determines the direction of the twisting motion that takes a molecule like butadiene into cyclobutene? The illustration below shows that there are two pathways, a "conrotatory" pathway, and a "disrotatory" pathway, that the molecule can take as it twists into a ring (the same two pathways also exist for the reverse reaction, when the ring bursts open).

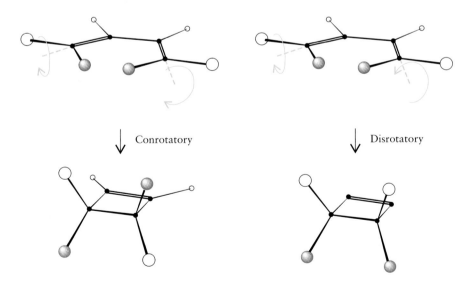

↓ Conrotatory   ↓ Disrotatory

There are two pathways by which a butadiene molecule (and, as shown here, its substituted versions) can change into cyclobutene. In the "conrotatory" path, the CH$_2$ groups rotate in the same sense, but in the "disrotatory" path they rotate in opposite senses from each other.

There are important consequences of the choice of path, for different products are obtained if the starting materials are slightly more elaborate than the materials we have been considering. For example, if there are different substituents in the starting materials, then as shown in the illustration, they behave like flags, and, depending on the path taken by the reaction, the final product will have the flags flying on the same side or on opposite sides of the ring. We shall see that which of the two paths is adopted (and hence which of the two products is obtained) depends on how the reaction is initiated, by heating or by absorption of light.

Butadiene and its analogs can be stored unchanged for long periods. This indicates that the ring-formation reaction is an activated process in the sense explained in Chapter 5. That is, as the molecule begins to twist into the shape it needs to close the ring, its energy rises, and only those molecules with enough energy to complete the twisting motion can go on to form products. The activation barrier stops the molecule twisting loosely into cyclobutene. It also stops cyclobutene untwisting loosely back into butadiene unless it has enough energy (from a collision, for instance).

A part of the strategy for analyzing the details of the reaction is to discover the origin of the block on the path. We shall find that the height of the barrier—the activation energy of the reaction—depends on the path, whether it is conrotatory or disrotatory, and hence one route may be faster than the other. We shall see that we can identify which route has a high activation barrier by tracing the changes to the molecular orbitals that take place as the molecule twists itself into a ring. If we find very high activation

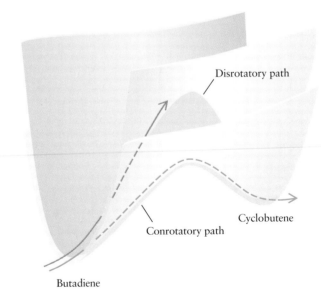

A path between reactant and product molecules that has a low activation barrier will allow a much faster reaction than a path for which the activation barrier is high. If one barrier is very high, the entire reaction course will follow the other pathway.

CHAPTER EIGHT

energy for a particular transformation, we may be able to conclude that the path is blocked and that the transformation does not occur. The key to understanding why the conrotatory and disrotatory paths have different activation barriers lies in the symmetry of the molecule. At root, then, we are seeking how the symmetry of a molecule determines its destiny.

To make progress, we need to draw on the material introduced in Chapter 2. We saw there that electrons occupy molecular orbitals built from the valence orbitals of the atoms. Each carbon atom in butadiene has four valence orbitals and contributes four electrons to the molecule. Three of the orbitals and three of the electrons from each atom are taken up with σ-bond formation (either to hydrogen atoms or to carbon atoms). That leaves each carbon atom to contribute one atomic orbital and one electron to the π molecular orbital system. We shall concentrate on these π orbitals because they are the seat of the changes that take place: the σ orbitals act as a flexible platform on which the more mobile π electrons play their parts.

There are four π molecular orbitals, each one being a combination of the available atomic orbitals but with different distributions of "color." The orbitals are shown on the left in the illustration on the next page. The four electrons provided by the carbon atoms pair and occupy the two orbitals of lowest energy; they should be thought of as spread over all four atoms, with their binding influence shared among them. When the reaction takes place and the ring forms, one lobe from the atomic orbitals on each of the two outermost atoms twists toward the other. The two lobes overlap, form a local σ orbital between the two end atoms of the chain, and hence pin the ends together into a ring. The molecular orbitals that can be built from the four atomic orbitals in cyclobutene are shown to the right in the illustration on the following page: two of the orbitals now correspond to bonding and antibonding σ orbitals between the atoms that are newly joined when the ring forms.

However, when we look at the reaction in more detail, it becomes necessary to note that the *symmetry* of the molecule changes as the ring closes. We have already seen that the symmetry of a molecule and its molecular orbitals plays a role in determining the outcome of a reaction. Now we have to be rather more sophisticated than in Chapter 7 because the orbital symmetries are more complicated, and a molecule of one symmetry changes into a molecule with a different symmetry as the reaction takes place. Moreover, because the orbital lobes of the molecule are like decorations that make the shape of the molecule more elaborate, we must distinguish between the symmetry of the molecule and the symmetries of the orbitals that ride on the atoms as the molecule twists from one shape to another.

We use two ideas to keep track of the changes that occur to orbitals as the molecule twists into a ring. The first is that we can identify an orbital from its symmetry, using the fact that each orbital can be classified as hav-

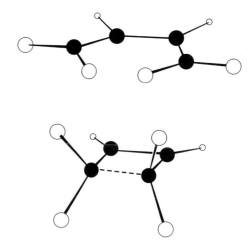

The σ-bond framework of the butadiene and cyclobutene molecules. We shall not consider these bonds explicitly, for they are largely unchanged by the reaction. However, the framework is slightly stretched and bent as the molecule changes into a ring, and so the σ bonds do contribute to the activation energy of the cyclization reaction.

Left: The four π molecular orbitals that can be formed from the carbon atomic orbitals. Their relative energies can be assessed from the number of internuclear nodes: as we saw in Chapter 2, the greater the number of internuclear nodes (where opposite colors meet), the higher the energy of the orbital; the lowest energy orbital is the one with no nodes (other than the one in the molecular plane). Each orbital can accommodate two electrons, and so 1π and 2π are occupied in the ground state of the molecule. Right: The four molecular orbitals in cyclobutene that can be formed from the four carbon atomic orbitals that contribute to the π orbitals in butadiene. Since σ orbitals between carbon atoms are stronger than π orbitals (and σ antibonding orbitals correspondingly more antibonding), the σ orbital and its antibonding partner lie at low energy and high energy respectively.

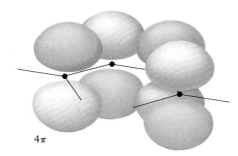

4π

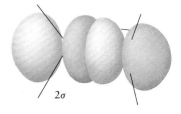

2σ

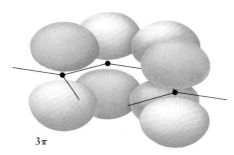

3π

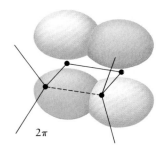

2π

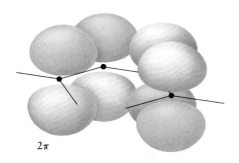

2π

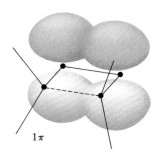

1π

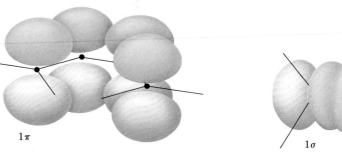

1π

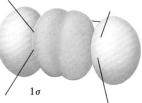

1σ

ing one of two types of symmetry (in a sense we shall shortly explain). The second is that an orbital cannot suddenly flip into a new symmetry type as the molecule twists from being a reactant to being a product. Therefore, if an orbital has a particular symmetry as a reactant, it will usually keep that symmetry all along the reaction path (that is, as the molecule twists into product), and it will emerge from the reaction with the same symmetry as it had initially.

The problem we immediately encounter in keeping track of the changes to orbitals is that, because the shape of the molecule is changing as it forms a ring, its symmetry is also changing, and so the symmetry of the orbitals also changes. However, we can follow the orbitals by noting that even though the reactant and product molecules have different shapes, certain symmetry features are common to both. For instance, as shown in the margin, both cyclobutene and butadiene share a twofold rotation axis and two mirror planes. It follows that we may be able to trace the fates of the orbitals in butadiene by identifying the features that are preserved throughout the course of the reaction, and use only these "conserved" features in the analysis. However, once we consider the reaction path, we discover that two of the shared symmetry features are lost as soon as the molecule starts to twist and are recovered only when the ring has fully formed. Thus, only one symmetry feature is retained throughout the reaction. Moreover, a *different* feature is retained on each of the two possible paths, the twofold axis for the conrotatory path and one of the mirror planes for the disrotatory path.

Once we have identified the conserved symmetry features, we can keep track of the changes that occur when the ring forms by identifying which orbital of the reactant becomes which orbital of the product. This is where the "colors" of the orbitals enter the picture. Colors allow us to classify orbitals by noting their behavior when the molecule is rotated or reflected: the *molecule* retains its original appearance under such a transformation, but the *orbital* might change color. Since classical physics is colorblind it could not deal with these features; we are now in the midst of the quantum mechanism of reactions.

First, some notation: If a rotation or a reflection of the molecular framework leaves the colors of an orbital unchanged, we say that the orbital is "symmetrical" with respect to the symmetry feature and label it S. If red and green are interchanged, we say that the orbital is "antisymmetrical" with respect to that feature and label it A. Consider the conrotatory path. The twofold rotation axis survives all along the path from reactant to product, and so we can use it to classify the orbitals as S or A at all points of the path. Now we can begin to track the orbitals from reactant to product, since we know that the S orbitals of the starting material become the S orbitals of the final product, and the A orbitals of the reactant become the A orbitals of the product.

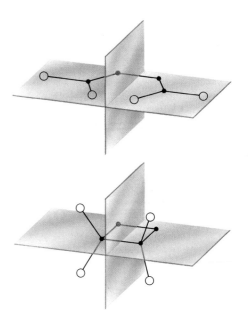

By a "symmetry feature" of a molecule is meant a transformation that leaves it apparently unchanged. A butadiene molecule (top) has two "mirror planes" in which one half of the molecule is an exact reflection of the other half. It also has a "twofold rotation axis," which means that it is left apparently unchanged when it is rotated by $360°/2 = 180°$ about the axis formed by the intersection of the mirror planes. Despite its different shape, a cyclobutene molecule (bottom) also has a twofold rotation axis and two mirror planes.

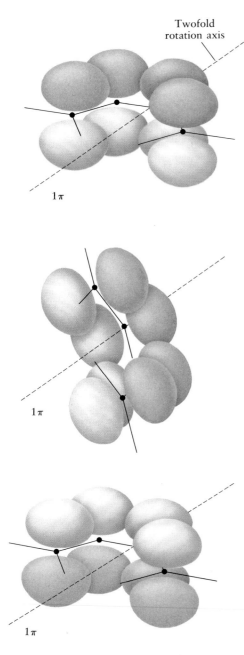

Twofold rotation axis

1π

1π

1π

The 1π orbital of butadiene is antisymmetric with respect to rotation of the molecule around the twofold axis, since the colors of the orbitals are reversed.

But first we come up against another crucial point. There are two S orbitals and two A orbitals on the reactants and the products, so we need to know which S orbital of the product each S orbital of the reactant becomes (and likewise for the A orbitals). The question is resolved by the "noncrossing rule" of quantum mechanics. This rules asserts that *orbitals of the same symmetry do not cross*. We can imagine the energy of one orbital rising and the other falling as the molecule twists into a ring. However, instead of the two orbitals reaching the same energy as the twisting reaches a certain angle and then crossing, so that the initially lower-energy orbital becomes the higher, the orbitals move apart in energy as the twisting continues. That is, before the two orbitals can cross, the initially rising orbital begins to fall in energy, while the energy of the previously falling orbital begins to rise. The effect is purely quantum mechanical, and arises from a kind of interference between the two orbitals. Later we shall need to note that the noncrossing rule applies strictly only when the twisting occurs very slowly. It applies to a good approximation to many of the molecular motions we shall consider, but if the molecule were suddenly to jerk from one shape to another, the electronic rearrangements that are needed for the noncrossing rule to apply might not have time to take place. For the time being, however, we shall suppose that the rule is obeyed strictly.

Since orbitals of the same symmetry cannot cross, each S orbital of the initial molecule turns into the S orbital that lies nearest it in energy when the molecule has become a ring; each A orbital behaves similarly. (There is nothing to stop S and A orbitals crossing, for they have different symmetries.) At last we have the information we need to draw a "correlation diagram" that shows how the orbitals of butadiene convert into the orbitals of cyclobutene as the end atoms twist into a ring. The correlation diagram for the conrotatory path is shown in the left of the illustration on the facing page.

Now consider the disrotatory path. Since the mirror plane is now the symmetry feature of the molecule that is preserved throughout the transformation from reactants to products, we keep track of the symmetries of the orbitals as the molecule twists into its ring form by concentrating on their properties with respect to this reflection. The molecular orbitals of butadiene are S, A, S, and A, respectively, and those of cyclobutene are S, S, A, and A, respectively. The noncrossing rule then leads to the correlation diagram shown on the right of the illustration.

Now we can see the difference between the two pathways. In a thermal reaction (one caused by heat), the initial molecule acquires the excess energy it needs to snap into a new orientation by collisions either with the surrounding medium or with other molecules in a gas. However, collisions, although energetic, are not *very* energetic, and cannot supply very much energy to a molecule. In particular, they usually cannot lift a molecule into

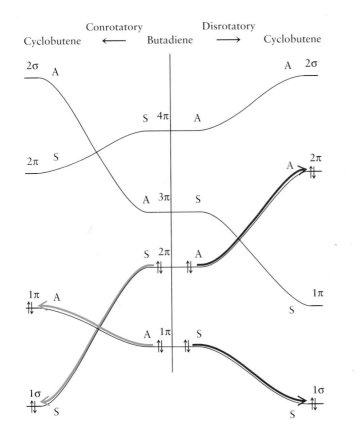

Conrotatory · · · · · Disrotatory

Cyclobutene ← Butadiene → Cyclobutene

The correlation diagram for the interconversion of molecular orbitals of butadiene and cyclobutene. The diagram summarizes how the energies of the orbitals change as the molecule is twisted into its ring form.

an electronically excited state, since to move electrons from one distribution into another requires large amounts of energy.

Initially, the two π molecular orbitals of lowest energy are occupied in butadiene. We write this "electron configuration," or orbital occupancy, as $1\pi^2 2\pi^2$ to signify that there are two electrons in the 1π orbital and two in the 2π orbital. We see from the correlation diagram that, for a conrotatory reaction path, this configuration changes smoothly into the configuration $1\sigma^2 1\pi^2$ of the product, in which two electrons occupy a σ orbital and two electrons occupy a π orbital: a π bond has been replaced by a σ bond and the activation energy of the reaction is minimal. On the other hand, under the disrotatory path, the electron configuration becomes the $1\sigma^2 2\pi^2$ configuration of the product. Although a π bond has been replaced by a σ bond, the other two electrons have been elevated into an orbital of high energy, and the product molecule is in a highly excited electronic state. The considerable energy needed to achieve this process is unavailable from collisions, and under thermal conditions it cannot occur. We have to conclude that in

Only if the end groups twist in a conrotatory sense does the ground state of a cyclobutene molecule become the ground state of butadiene (with electron configuration $1\pi^2 2\pi^2$); this is shown by the green lines. The alternative, disrotatory ring-opening pathway leads to the electronically excited state $1\pi^2 3\pi^2$ of butadiene, as shown by the red lines, and this pathway is too energy-demanding to be feasible.

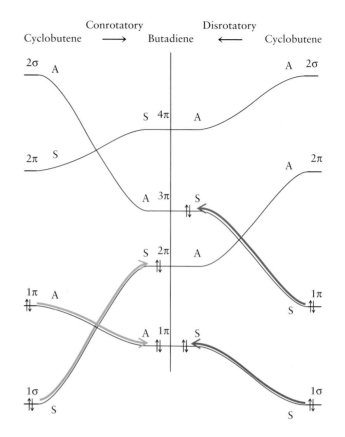

the thermally induced cyclization of butadiene, the product is that of the *con*rotatory path. Likewise, if we read the correlation diagram in reverse, starting at the $1\sigma^2 1\pi^2$ ground state of cyclobutene and seeing how that state evolves as the ring opens, then we can conclude that in the thermally induced ring-opening of cyclobutene the molecule also twists in a conrotatory sense.

Whether the ring-closing or ring-opening reaction will be conrotatory or disrotatory depends on the number of electrons in the π orbitals of the starting material. Take for example, the analogous reaction below:

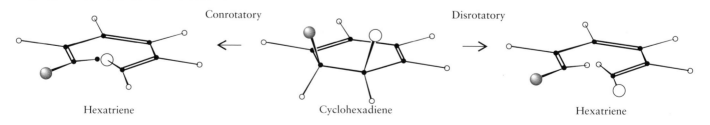

In this reaction, there are six electrons in the molecular ground state and hence three occupied π orbitals. The same analysis as before leads to the conclusion that the thermally feasible reaction path is the one that corresponds to *dis*rotatory ring-opening, with the end groups rotating in opposite directions. Hence, the product will be the structure that arises from this mode and not the conrotatory scission shown by cyclobutene.

There is a general prediction for thermally induced electrocyclic reactions that, as the number of electrons in the π system changes along the series 4,6, . . . , the reaction will alternate conrotatory, disrotatory, ... This prediction is one of the celebrated "Woodward-Hoffmann rules" that help to rationalize a great swathe of organic chemistry.

CYCLOADDITION REACTIONS

We can explore by similar means the pathway adopted in a cycloaddition reaction to see how symmetry strangles some routes and facilitates others. We shall consider two contrasting examples of cycloaddition: the union of two molecules, or "dimerization," of ethylene to form cyclobutane,

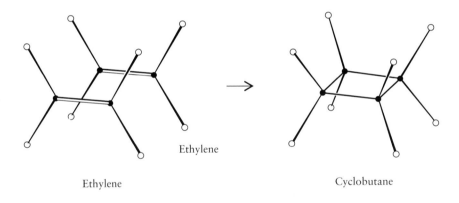

Ethylene

Ethylene

Cyclobutane

which under thermal conditions proceeds very slowly, and the very much faster Diels-Alder addition of ethylene to butadiene that was mentioned at the start of the chapter. The fact that the thermal dimerization of ethylene is much slower than the thermally induced Diels-Alder addition is of considerable commercial significance, for it means that ethylene can be preserved for indefinite periods until it is to be used to synthesize the polymers that are so important to modern life. If dimerization were fast, industry would be unable to produce polymers such as polyethylene, polystyrene, and polyvinyl chloride.

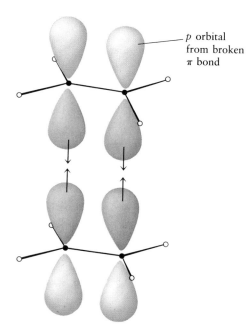

p orbital from broken *π* bond

When two ethylene molecules approach face to face, there appears to be nothing in principle to stop the π bonds from opening and the atomic orbitals from overlapping to form σ bonds that link the molecules into a four-membered ring.

The reason why ethylene does not dimerize when two molecules collide in a heated container is rooted in the quantum mechanical consequences of symmetry; it is symmetry that preserves the raw product from annihilation. To see how this is so, consider the face-to-face approach of two ethylene molecules shown in the margin. In order for molecules to react, the *π* component of the carbon–carbon double bond must be broken, making available a *p* orbital on each carbon atom. As the two molecules approach, each *p* orbital will meet another head-on to form a *σ* bond with the other molecule, so completing a four-membered ring.

So far, there would not appear to be any reason why the dimerization reaction should not be successful. However, now consider the symmetry of the molecules as they approach each other, and the colors of the orbitals. The two mirror planes shown in the illustration below are present in the initial encounter, while the reaction is in progress, and in the final product. It follows that we can classify the orbitals with respect to these *two* mirror planes, and that we can construct the correlation diagram by considering the S or A symmetry of the orbitals with respect to *both* planes. In the illustration at the top of the facing page, we show how matching *π* orbitals pair with each other as the two ethylene molecules approach, and we indicate the joint classification of each pair. The *σ* orbitals that the approaching molecules can form with each other can also be classified as A or S with respect to each plane, as shown in the illustration at the bottom.

To evaluate the feasibility of the reaction, we need to know the relative energies of the molecular orbitals. As usual, the relative order of energies can be judged by assessing the number and importance of internuclear

A pair of ethylene molecules (left) meeting face-to-face is symmetrical with respect to two perpendicular mirror planes. So is their product, a cyclobutane molecule (right). The two mirror planes are present throughout the course of the reaction.

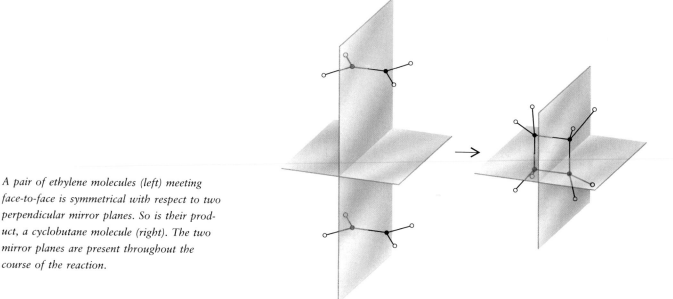

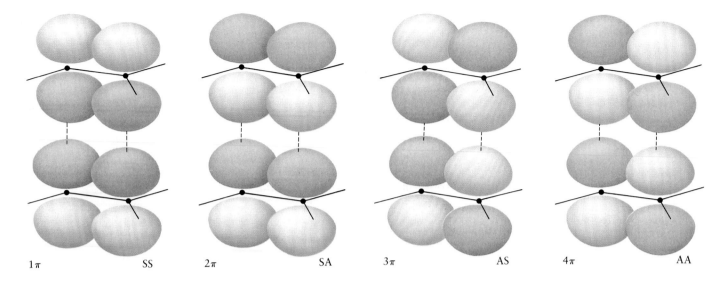

| 1π | SS | 2π | SA | 3π | AS | 4π | AA |

The symmetry of the pairs of π orbitals on two ethylene molecules approaching each other face-to-face can be classified with respect to the two mirror planes shown in the preceding illustration. SA, for instance, indicates that the orbital is symmetrical with respect to the vertical mirror plane and antisymmetrical with respect to the horizontal one.

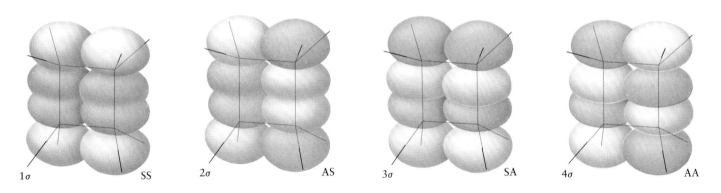

| 1σ | SS | 2σ | AS | 3σ | SA | 4σ | AA |

The classification of the new σ bonding and antibonding orbitals that are formed by overlap of the p orbitals that were initially used for π-bond formation in the two ethylene molecules. (When assessing the relative energies of the molecular orbitals, we note that there is little overlap between orbitals that are broadside across the ring: the principal overlap is that between orbitals that are pointing directly at each other.)

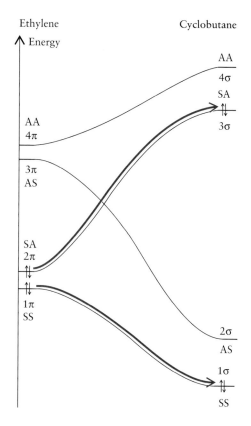

Ethylene   Cyclobutane

Energy

AA

4σ

SA

AA
4π

3σ

3π
AS

SA
2π

AS
1π
SS

2σ

AS
1σ

SS

The energy levels of the π orbitals of two ethylene molecules (left) and of the σ orbitals they form in cyclobutane (right). The correlation between them is also shown.

nodes in each orbital. The correlation diagram is now completed by connecting orbitals of the same symmetry without them crossing. It is clear from the correlation diagram shown on this page that at the end of the reaction the dimerization process leads to a cyclobutane molecule in an excited electronic state. This process requires so much energy that it cannot be induced merely by heating the reactants and expecting the ethylene molecules to smash together sufficiently energetically; hence the ethylene survives for indefinite periods. Thus, dimerization is forbidden by the symmetry of the orbitals and the individuality of the molecules is preserved.

Now let us see why the Diels-Alder addition of ethylene to butadiene is so much more facile. Once again, we consider the face-to-face approach of the two molecules, which preserves the mirror plane shown in the illustration below. In the course of the reaction, two new σ bonds are formed at the expense of two π bonds (and one π bond moves to a new location).

The correlation diagram for the addition reaction, shown on the facing page, is constructed in the usual way, and we can trace the evolution of the bonding electron pairs in the three lowest orbitals of the initial molecules as the reactants evolve into products. The obvious feature is that the ground-state electron configuration of the reactant correlates with the ground-state electron configuration of the product (in which the three electron pairs continue to occupy the lowest three orbitals). As a result, the cyclohexene molecule is formed in its ground electronic state. The smooth transition from reactants to products involves no excitation of electrons, and its activation energy can be expected to be sufficiently low for the reaction to be feasible thermally. This conclusion is in full accord with the observed rate of the Diels-Alder addition; it is a thermally *allowed* reaction.

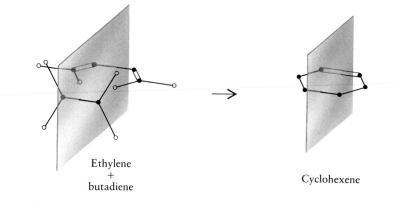

Ethylene
+
butadiene

Cyclohexene

The mirror plane that is common to ethylene and butadiene in a Diels-Alder reaction.

PHOTOCHEMICAL VERSUS THERMAL REACTIONS

A reaction that is thermally forbidden may become viable when electrons are excited into different orbitals by the absorption of photons of light. Such reactions occur not merely, as might have been thought, because more energy is available in the starting materials after they have absorbed the energy carried by a photon, but also because reaction pathways become opened by modification of the symmetry arguments that we have used so far. Light renders pathways accessible not only by providing greater energy *but also* by rendering the forbidden allowed. After the absorption of light, one electron will occupy a different orbital, and the correlation diagram will have different consequences, which may include newly opened reaction pathways. We can begin to appreciate the exquisite control that may be had over matter at a molecular level, for a thermal reaction may lead to the rotation of the groups in one direction, whereas a photochemical reaction starting with the same materials may result in rotation in the opposite sense and hence give a different product.

Consider the photochemical ring closure of butadiene from the point at which the absorption of a photon has led to the excitation of a single electron, from the 2π orbital to the 3π orbital. When we follow the disrotatory pathway (shown in the correlation diagram on the following page), we find that the excited electron configuration of the butadiene molecule crosses into an excited electron configuration of the cyclobutene molecule having an energy similar to that of the initial molecule, whereas the conrotatory path involves a significant increase in energy. Hence, in contrast to the thermal electrocyclic reaction, the disrotatory path is open for the photochemical reaction and the conrotatory is closed.

The photochemical ring closure of butadiene is in fact known to occur by the disrotatory path. Nevertheless, there are some complications (as in most photochemical reactions), and cyclobutene is produced in its *ground* electronic state, not its excited state as expected from our discussion. To resolve this discrepancy we must climb to our final plane of understanding.

THE CORRELATION OF OVERALL STATES

The problem with orbital diagrams is that they focus attention on individual *orbitals* rather than the collective behavior of all the electrons in the molecule, which is the true determinant of the course of the reaction. We must see how to consider the joint behavior of collections of electrons.

Suppose that in a molecule one electron occupies an orbital that is S with respect to a rotation or a reflection, and another electron occupies an

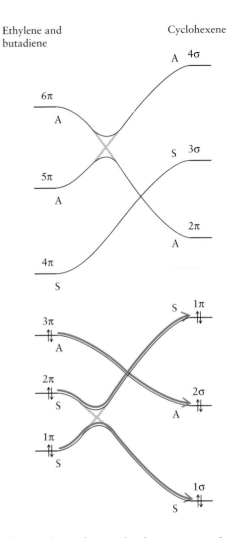

Ethylene and butadiene — Cyclohexene

The correlation diagram for the conversion of the π orbitals of ethylene and butadiene to the orbitals of cyclohexene in a Diels-Alder reaction. Note how the $1\pi^2 2\pi^2 3\pi^2$ ground-state configuration of the starting materials becomes the $1\sigma^2 2\sigma^2 1\pi^2$ ground-state configuration of the product.

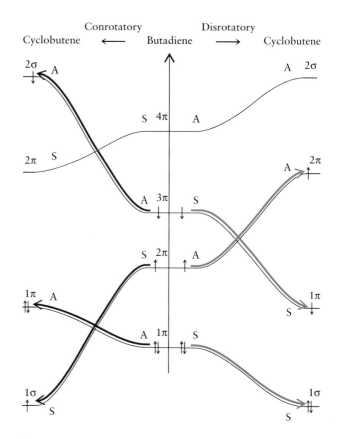

The correlation diagram for the photochemically induced conversion of butadiene into cyclobutene. The absorption of a photon raises a 2π electron into a 3π orbital. The conrotatory twisting of the molecule now results in the formation of product in a highly excited state, and so is forbidden.

Conrotatory   Disrotatory

Cyclobutene $\longleftarrow$ Butadiene $\longrightarrow$ Cyclobutene

orbital that is also S. Then the *overall* effect of the rotation or reflection is S. If both occupied orbitals are A, then the *overall* effect is still S. To appreciate why, we need to note that the colors of the orbital represent the signs of a wave, and if both orbitals change sign the overall effect is equivalent to no change of sign. A similar argument applies when one orbital is S and the other is A, but now the joint classification is A because one orbital changes sign but the other does not. We summarize these remarks by the rules

$$S \times S = S \qquad A \times A = S \qquad S \times A = A$$

The same rules are used if there are more than two electrons present. For instance, if there are three singly occupied A orbitals, the overall symmetry classification is

$$A \times A \times A = A \times (A \times A) = A \times S = A$$

When all the orbitals are doubly occupied, even numbers of S and A are multiplied together to give S overall. For this reason, almost all molecular ground states have S symmetry.

To see how these rules work in practice, we consider the first few excited electronic states of butadiene and cyclobutene, and identify their symmetries in terms of the disrotatory path. Since the disrotatory path preserves the single mirror plane, the orbitals are classified A or S with respect to that plane. The first excited electron state of butadiene has the configuration $1\pi^2 2\pi^1 3\pi^1$, with one electron elevated from the 2π orbital (an A orbital) up into the 3π orbital (an S orbital). Its overall symmetry is therefore

$$\underset{1\pi^2}{(S \times S)} \times \underset{2\pi^1}{A} \times \underset{3\pi^1}{S} = A$$

In the next higher configuration of butadiene, both the 2π electrons are promoted up into the 3π orbital, so its configuration is $1\pi^2 3\pi^2$. Since all the occupied orbitals are S, the overall symmetry of this state is

$$(S \times S) \times (S \times S) = S$$

Now we consider the overall orbital symmetries of the product molecule, cyclobutene. We know the individual classification of the orbitals with respect to the preserved mirror plane, and can use the multiplication rules to work out the overall symmetries for the ground state $(1\sigma^2 1\pi^2)$ and for the first couple of excited states, in which electrons are raised from the 1π orbital up into a 2π orbital to give the configurations $1\sigma^2 1\pi^1 2\pi^1$ (one electron raised) and, considerably higher in energy, $1\sigma^2 2\pi^2$ (both electrons raised). The symmetries of the resulting states are set out on the right of the illustration.

We can construct the first approximation to the state correlation diagram by referring to the orbital correlation diagram that we constructed earlier, shown on page 191. There we saw that the butadiene orbitals 1π and 2π become the cyclobutene orbitals 1σ and 2π; it follows that the butadiene configuration $1\pi^2 2\pi^2$ S becomes the cyclobutene configuration $1\sigma^2 2\pi^2$ S, and we draw a line connecting the two states. All the other first-order approximations to the correlation diagram lines may be drawn similarly, by noting the orbital correlations from the earlier illustration, and drawing the connecting lines.

The key point is that there are *two* S states that—at this stage—appear to cross: one state begins in the ground state of butadiene and ends in the highest excited state of cyclobutene; the other begins in the highest excited state of butadiene and ends in the ground state of cyclobutene. However, just as the noncrossing rule forbids orbitals of the same energy to cross, so it also forbids *states* of the same symmetry to cross. Once this rule is taken into account, the correlations follow the black lines in the illustration.

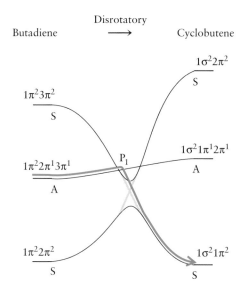

Butadiene   Disrotatory $\longrightarrow$ Cyclobutene

$1\sigma^2 2\pi^2$ S

$1\pi^2 3\pi^2$ S

$1\pi^2 2\pi^1 3\pi^1$ A P_1 $1\sigma^2 1\pi^1 2\pi^1$ A

$1\pi^2 2\pi^2$ S $1\sigma^2 1\pi^2$ S

The state correlation diagram for the conversion of butadiene into cyclobutene for the ground state and first two excited states of the butadiene molecule.

Now let us see where we have arrived. The correlation diagram shows that the ground state of butadiene (lower left) becomes the ground state of cyclobutene (lower right). However, the correlation diagram shows that there is an Everest-like peak on the reaction path, and therefore that the disrotatory mode of twisting is blocked: the initial twisting phase of the interconversion requires the molecule to acquire so much antibonding character before the motion converts that orbital into one with more bonding character that the molecule is locked into its initial form. The same is true of ring scission: the initial twisting motion of the molecule as it struggles to burst its ring requires the acquisition of too much antibonding character for scission to be feasible as the result of a collision. The disrotatory path is blocked by the molecule's symmetry.

Now consider the photochemical reaction, and trace what happens to the molecule when an electron is boosted up into another orbital. In a typical photochemical process, the absorption of a photon of light promotes an electron from a 2π orbital up into a 3π orbital, and the molecule settles down for a brief career as an excited A state. As the molecule twists toward becoming cyclobutene, the uppermost S state decreases in energy as one of its π orbitals is converted into a σ orbital and at a certain angle of twist it has the same energy as the A state. At that point, the electron distribution is at its most responsive to tiny influences, and it is very easy for the electrons to adjust their positions: it is a general rule of quantum mechanics that states of equal energy are sloppy, and can easily be transformed from one to the other. Thus, if a nucleus moves a little too fast for an electron to follow, it might result in the electron distribution being pushed into a new arrangement. In the present case, at the instant the S and A states have the same energy, the flick of a nuclear motion can result in the molecule switching on to the S curve and then plunging down past the crossing point and becoming ground-state cyclobutene. The interconversion has been successful: photochemical excitation has resulted in the formation of a cyclobutene molecule by a *dis*rotatory path, the path that was closed to reactions that took place by collision. Now symmetry considerations have opened a path that previously they had blocked.

PHOTOCHEMICAL BLOCKING

We have seen that the absorption of a photon can open a reaction pathway. More surprising, perhaps, is that the absorption of a photon can also block a pathway that was previously open. (This shows that the effect of photon absorption is not merely the acquisition of a greater energy by the molecule.) The Diels-Alder ethylene-butadiene addition reaction is one example of a reaction that is thermally allowed but photochemically forbidden.

We can see how a photon can block an otherwise perfectly feasible reaction by considering the correlation diagram. As usual, we use the individual orbital correlation diagram to build the framework of the state correlation diagram, and then apply the noncrossing rule to arrive at the diagram shown in the margin. The most obvious feature of this diagram is the absence of any energy barrier between the two ground-state configurations; hence the reaction is thermally allowed. (The small activation energy that exists in practice stems from the adjustments made to the σ framework of the molecule as it twists into its new shape.) Now suppose a photon plunges into the approaching pair of molecules and excites an electron from a 3π orbital into a 4π orbital; now the pair has the configuration $1\pi^2 2\pi^2 3\pi^1 4\pi^1$, an overall A state. According to the correlation of *orbitals*, this configuration would change into the highly excited configuration $1\sigma^2 2\sigma^1 1\pi^2 3\sigma^1$ of the addition product. However, as the ethylene and butadiene molecules merge and the A state rises in energy, it approaches another A state that stems from an even more highly excited configuration, and which falls in energy as the molecules merge. The noncrossing rule forbids their intersection, and so the lower A state of the pair of molecules evolves instead into the lower A state of the product. As a result of its initial rise in energy and subsequent fall as the electron distribution changes, there is a substantial hump in the correlation diagram, and the photochemically initiated process is forbidden even though the molecule possesses more energy than in its ground state.

We have come a long way from Faraday's interests as we have pursued the carbon atom into these reactions. Nevertheless, if the carbon atom released from the candle were to make the transition into the biosphere, it would be open to transformations such as these. The lesson of this chapter, though, is not so much that carbon is rich in its properties—that should be plain enough by now—but that to *understand* its behavior fully we must let slip our grasp on the comfortable, familiar world of classical concepts, and admit that quantum mechanics is the regulator of change. In particular, through the workings of quantum mechanics, the possession by a molecule of a particular symmetry can determine its chemical destiny.

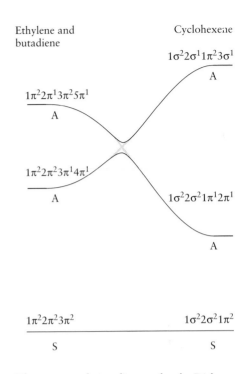

Ethylene and butadiene

Cyclohexene

$1\sigma^2 2\sigma^1 1\pi^2 3\sigma^1$

A

$1\pi^2 2\pi^1 3\pi^2 5\pi^1$

A

$1\pi^2 2\pi^2 3\pi^1 4\pi^1$

A

$1\sigma^2 2\sigma^2 1\pi^1 2\pi^1$

A

$1\pi^2 2\pi^2 3\pi^2$

S

$1\sigma^2 2\sigma^2 1\pi^2$

S

The state correlation diagram for the Diels-Alder addition of ethylene to butadiene. The transformation from ground state to ground state can occur readily as there is no activation barrier (if we ignore the changes to the σ framework). However, a high barrier blocks the reaction of an electronically excited state.

LIGHT
AND LIFE | 9

Green vegetation uses the energy of solar radiation to build carbohydrates through the complex sequence of chemical reactions known as photosynthesis.

I have also here some candles sent me by Mr Pearsall,

which are ornamented with designs upon them, so that,

as they burn, they have, as it were, a glowing sun

above, and a bouquet of flowers beneath.

MICHAEL FARADAY, LECTURE 1

One of the great journeys of the world is the one in which carbon steps from an inorganic existence, as carbon dioxide, to an organic one, as glucose. This step is photosynthesis, in which green plants utilize the energy provided by the sun and, by plucking carbon dioxide from the air, convert it into carbohydrate. Photosynthesis is the reaction that lies at the head of the food chain, for in capturing the energy of the sun and trapping it as an elaborate form of matter, it generates a primary fuel. Without the reactions of photosynthesis, the earth would be a wet, warm rock populated by primitive organisms able to harvest only very inefficiently the avalanche of energy hurled at us from the sun. With the reactions of photosynthesis, it is a vibrant green globe thronged with life that is supported by the pyramid of consumption that projects upward from its carbohydrate base and culminates currently in carnivores. Through photosynthesis, the sun coils the spring of all our activities.

It was ultimately through photosynthesis that Faraday, more cautiously than Prometheus, snatched power from the sun. The chemical prehistory of his candle included a step in which a carbon dioxide molecule was captured from the air and built by photosynthesis into a carbohydrate that later fed the bee that formed the wax. When Faraday lit his candle, he let slip the power that had been captured—perhaps years previously—from the sun and stored meanwhile in the molecules that photosynthesis had helped to form.

Whereas some chemical reactions are stimulated by light, there are others that produce light. Some do so indirectly, as when the combustion of a fuel in a power station, or the redox reaction in a battery, sends electricity to a lamp. Others emit light as a consequence of their exothermicity, as when the candlewax burns, leaving atoms or molecules and their fragments in an excited state, as in pyrotechnics. Yet other reactions—which we shall consider in detail in this chapter—generate light *directly* in the process called chemiluminescence.

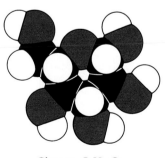

Glucose, $C_6H_{12}O_6$

In the process of photosynthesis, light from the sun is captured by the molecules in green leaves and used to convert carbon dioxide and water into carbohydrate. The by-product is oxygen, which can be seen here as tiny bubbles leaving the leaves of an underwater plant.

Photosynthesis represents the apotheosis of the events that we have been considering. Until now, we have introduced simple molecular events, and have focused on single steps that may contribute to reactions. Now we move to the stage where we leave the details and consider the global consequences of their contribution, as they are exemplified by the intricate interplay of the steps in the photosynthesis reaction. In this final chapter, the individual notes of chemical reactions are still present, but we stand further back and, instead, listen to the symphony of change.

THE CHEMICAL GENERATION OF LIGHT

Nature stumbled on chemiluminescence long before chemists trod the earth, for it is responsible for the glow of fireflies, glowworms, jellyfish, and rotting fish. Some bacteria are chemiluminescent, as is the fungus *Clitocybe illudens,* the aptly named "Jack-o-lantern."

The trick that nature and chemists use to conjure light from a reaction is to contrive to form products in an energetically excited state that subsequently discard the excess energy as radiation, instead of heat. That the whole world is not aglow with radiation is a consequence of a competition

The fungus Pulverboletus ravenelii *("bolete mushrooms") is chemiluminescent, glowing in the dark as a result of light-producing reactions.*

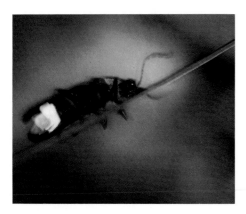

Fireflies are beetles of the family Lampyridae. *Their larvae are also luminescent, perhaps as a warning to predators that they are toxic. Adult fireflies flash to seek a mate; some are carnivores and simulate the flashing pattern of other species to attract them as prey.*

between the discarding of energy as radiation and as heat. The products of most reactions are in such intimate contact with their surroundings that any excitation is quickly transferred to the neighboring molecules in the form of thermal motion. However, there are some reactions for which the contact is so weak that the excited state survives long enough for the relatively slow business of squeezing out a photon to occur.

Natural but nonbiological sources of "cold" light include the aurora. This luminous display forms when the impact of energetic particles—mostly electrons and protons—from the sun ionizes nitrogen and oxygen molecules high in the stratosphere. The air there is so rarefied that collisions between molecules are infrequent and the excited molecules have time to generate photons. In a typical process, a nitrogen molecule, N_2, is ionized by electron impact:

$$N_2 + e^{-*} \longrightarrow N_2^{+*} + 2e^-$$

where the asterisk denotes a high-energy species. The N_2^+ ion can capture a less energetic electron and form an electronically excited state of the neutral molecule that can discard its energy as ultraviolet radiation and as violet and blue light. Alternatively, an oxygen atom can be excited by electron impact,

$$O + e^{-*} \longrightarrow O^* + e^-$$

and the excited atom can discard light of a whitish-green or crimson hue. Reactions that generate auroras are simulated on a smaller scale here on earth when an electric switch is thrown, and a spark is generated at the touch of the contacts. The flash of green that is sometimes seen in these circumstances emanates from ionized O atoms produced by the dissociation of O_2 molecules.

The variety of chemiluminescence known as bioluminescence, because it is produced by living organisms, typically involves a light emitting molecule, luciferin, and an enzyme, luciferase. Luciferins are variable between species, but the firefly luciferin is typical. Under the action of luciferase, oxygen, and the universal local power store ATP (adenosine triphosphate), the luciferin molecule loses an OH group and is converted into an electronically excited state of the molecule shown below. This molecule discards a green photon that we perceive as the firefly's fire.

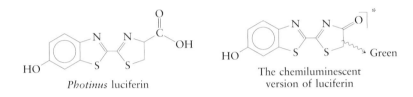

Photinus luciferin

The chemiluminescent version of luciferin

Bioluminescence is highly efficient in the conversion of chemical energy into radiation, for in some species only one percent of the energy appears as heat. A lower conversion factor is achieved in the laboratory generation of chemiluminescence, but even so the processes are reasonably efficient. One typical laboratory demonstration of chemiluminescence is based on the reaction of the compound luminol with hydrogen peroxide, H_2O_2, (in the presence of a catalyst, $[Fe(CN)_6]^{3-}$, which can participate in electron transfer), which produces an ion in an electronically excited state.

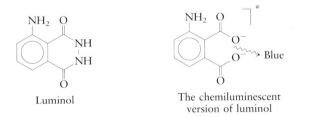

Luminol

The chemiluminescent version of luminol

The lightsticks that are sometimes used for pleasure or to provide light in an emergency are powered by reactions that work on principles similar to those we have described, but make use of an additional step. They typically consist of an outer plastic tube containing a solution of a fluorescent mole-

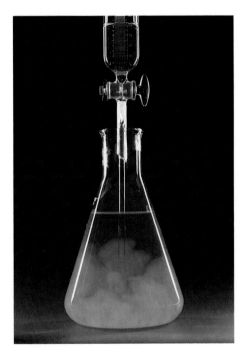

The chemiluminescent reaction between luminol and hydrogen peroxide forms a product in an excited electronic state that discards its excess energy as a blue photon, and the reaction generates an intense blue light.

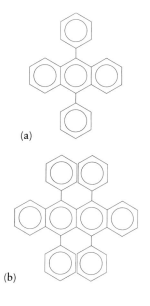

(a)

(b)

cule and an ester of oxalic acid, HOOC—COOH. Inside the lightstick itself is another tube containing hydrogen peroxide, H_2O_2, and a catalyst. When the inner tube is broken, the hydrogen peroxide reacts with the oxalic acid ester, to produce a cyclic compound:

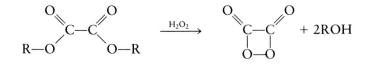

The highly strained four-member ring releases its energy to the fluorescent molecule, which is electronically excited in the process and then discards its excess energy as a photon. The fluorescer may take a variety of forms, including the molecule (a), which generates blue fluorescence, and (b), which generates orange.

THE CHEMICAL GENERATION OF COHERENT LIGHT

The light sources that we have been considering so far generate a random spray of photons, rather like the illumination from an incandescent lamp. However, chemists now know how to bleed off the energy released in chemical reactions as a beam of "coherent" radiation, radiation in which the waves produced by different molecules are all in step with each other rather than being generated at random. The reaction we are about to describe is the basis of the "chemical laser," and is a component of whatever remains of the extravaganza that formed the Strategic Defense Initiative.

There are two essential features of the action of a chemical laser. One is a chemiluminescent reaction that makes available photons. The second feature ensures that those photons are produced in a surge of radiation instead of spraying out from the reaction vessel at random (as from a firefly or a lightstick). Accomplishing the latter is largely a matter of finding the right design for the cavity in which the light is generated. As far as I know, there is no natural, biological version of a laser. However, since the physical character of the organs in which bioluminescence occurs is widely variable—in some species the organ is behind a tiny lens and is backed with reflective material that turns it into a little searchlight—it is possibly only a matter of time before a species evolves a minute natural laser for courtship, signaling, or defense.

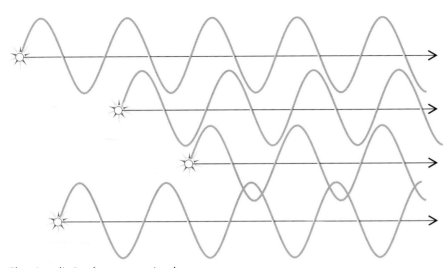

Chaotic radiation from conventional source

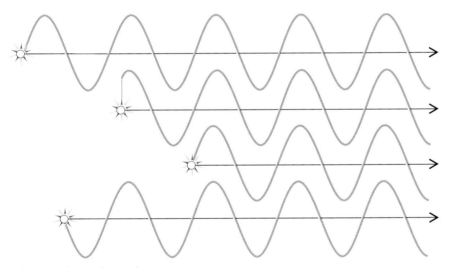

Coherent radiation from coherent source

The most highly developed chemical laser is the hydrogen fluoride laser. The essential step in the chemiluminescent reaction is the combination of hydrogen and fluorine atoms to produce hydrogen fluoride molecules that are vibrating vigorously:

$$H + F \longrightarrow HF^*$$

The hydrogen and fluorine atoms may be generated in a variety of ways. One is simply to mix hydrogen and fluorine gases and to rely on the radical reaction that ensues. Alternatively, an electric discharge can be passed through a gas such as sulfur hexafluoride, SF_6, consisting of molecules that are rich in fluorine atoms.

The vibrationally excited HF molecules are pumped rapidly into the laser vessel, which has mirrored walls. There the molecules are bombarded by a strong flux of photons that stimulates each molecule to emit its vibrational energy as a photon. These photons contribute to the radiation in the cavity, which lies in the infrared region of the spectrum. The HF molecules are now vibrationally dead and must be pumped away and vented into space lest they absorb the photons. Much of the radiation is allowed to escape in an intense beam, while some is trapped within the cavity and helps to stimulate the newly arriving HF^* molecules to discard their excess energy. Thus, so long as the supply of hydrogen and fluorine atoms is sustained, the reaction generates a beam of coherent radiation.

PHOTOSYNTHESIS

By virtue of the energy a photon can bring, light drives the most widespread of all chemical reactions on earth. Photosynthesis operates on a colossal scale. Each year, a hundred trillion kilograms of carbon are captured from the air and turned into carbohydrate. The energy stored in this way each year is about 10^{18} kilojoules, which is about 30 times the global consumption of energy. Joseph Priestley, the English Presbyterian minister, discoverer of oxygen, and inventor of soda water, first identified (in 1771) the ability of green plants to harvest carbon dioxide. Priestley recognized another of the crucial contributions of this reaction to our habitat when he identified oxygen as a by-product of photosynthesis and realized that a plant "could restore air which has been injured by the burning of candles." Today we recognize the global importance of photosynthesis to the cleansing of our atmosphere of the injuries that we inflict on it now that we have moved beyond mere candles.

Yet photosynthesis was not always the method by which solar energy was captured, and the stumbling into it by organisms had a greater environmental impact than any of humanity's current puny meddlings. Before the onset of photosynthesis there was little oxygen in the atmosphere—what there was had been released from the earth's crust and the impact of solar radiation on the water that had poured out of volcanoes. But once photosynthesis began, there was a sudden outpouring of its by-product, oxygen.

Almost all the oxygen in our atmosphere is the polluting effluent (from the viewpoint of the dominant lifeforms of the time) of the biological power stations found in green vegetation. The Great Oxygenation of the atmosphere brought in its train the possibility that life could lift up its roots and become mobile; thus animals emerged. Since animals hunt and are hunted, they must be able to respond immediately to tactical variability, and as a result intelligence evolved. Thus not only do we persist on account of photosynthesis, but our origin can be traced back to the onset of that reaction.

To understand more fully the events that accompany photosynthesis, I would like you to accompany me (in your mind's eye) toward and then into a leaf in search of the reactions within. Approaching from above we plunge through the cuticle that acts as a barrier to the uncontrolled evaporation of water and enter the mesophyll, the leaf's central cellular structure. Once inside the mesophyll we find ourselves in one of its numerous large intercellular spaces (for this is a factory where the supply of air is all-important) and surrounded by cells. In the dark green gloom, we can make out two main varieties of cell: some are columnar and others are irregular. Biologists have named the former the palisade parenchyma and the latter the spongy parenchyma (the parenchymatous cells are the most common kind of cells in plants). The former occur in greater abundance on the upper side of the leaf and contain the greater abundance of chloroplasts, the centers of photosynthetic activity. We might just be able to catch a glimpse of the external world

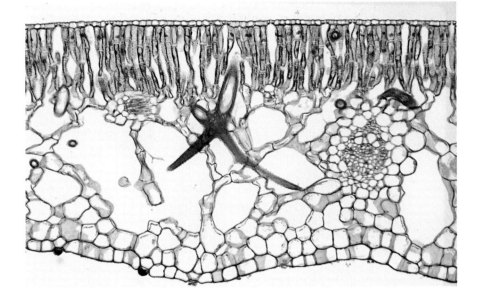

A cross section of a leaf of a water lily (Nymphaea odorata) has openings to the atmosphere (stomata) only on the upper surface. The palisade parenchyma are the vertical cells that lie under the upper epidermis, and the spongy parenchyma are the cells below.

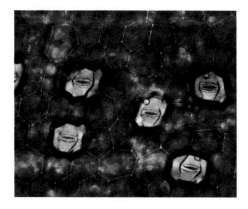

The guard cells, the epidermal cells, and the stomata of a wandering Jew plant. The plant exchanges gases with the atmosphere through the stomata. (The random arrangement of stomata on the leaf is typical of dicots; the stomata of monocots are more regularly distributed.)

through the leaf's stomata, for it is through these holes that the exchange of gases with the atmosphere takes place. The stomata are most numerous in plants that inhabit arid regions, enabling gases to exchange more rapidly with the surroundings during the brief periods when water becomes available and photosynthesis can proceed.

A chloroplast is the lair of our reaction. It is a sausage-shaped organelle that contains the thylakoids, the flattened, sacklike regions (thylakoid is from the Greek words for sacklike) that implement photosynthesis. The precise location of the initial light-driven steps of the photosynthetic reactions is in the tiny bodies that pepper the thylakoid membrane. The building work, the formation of carbohydrates from carbon dioxide, where carbon makes its auspicious transition, takes place in the medium surrounding the thylakoids, which is called the stroma (the Greek word for "spread out"). The thylakoid membrane is the site of the dynamos that drive the reactions taking place around it in the stroma.

When we move closer to the thylakoid membrane we see that the tiny dots that lie scattered over its surface are highly organized assemblies of molecules. Two of the most important types of assemblies are called photosystems. These systems are like miniature factories where the energy of the incoming light is trapped in matter. In them are hundreds of chlorophyll molecules that act as the primary receptors of the light, but they also include many other species that contribute to the transformation of the captured energy into a form that can be used to convert carbon dioxide into carbohydrate.

Chlorophyll is green: that accounts for the appearance of healthy vegetation and for the green gloom in which, in our imagination, we stand. However, packed into the photosystems are other molecules that give glints of different hue. Among them are the less fragile yellow, orange, and red carotene molecules that survive when the leaf begins to die and that are

The surface of the thylakoid is the location of the reactions of photosynthesis. The thylakoids are stacked into connected regions called grana.

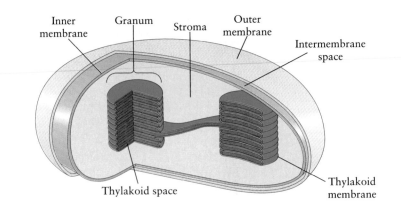

CHAPTER NINE

responsible for the colors of leaves in the fall. Their function during the lifetime of the leaf, when their color is masked by the intensely absorbing chlorophyll, is to contribute to the energy collection process and to protect the leaf from damage by the radiation to which it is exposed at the peak of the day.

Such is the general structure of the leaf, from the biological entity that bends in the wind to the molecular entities that reside in its chloroplast and that are responsible for its function. Now we must see how it works.

THE REACTIONS IN THE LEAF

The ratio of the numbers of carbon, hydrogen, and oxygen atoms in a carbohydrate molecule is typically $1:2:1$; this composition can be represented by the chemical formula CH_2O. It is this conjunction of the chemical formulas of carbon and water that gives rise to the name "carbo-hydrate," for the formula suggests a "hydrate" of carbon. However, carbohydrates are actually intricate assemblies of carbon, hydrogen, and oxygen atoms that merely happen, quite often, to have formulas that are multiples of CH_2O. For example, the carbohydrate that we shall consider as representative of them all is glucose, $C_6H_{12}O_6$, which was shown earlier. Cellulose and starch, the two main products of photosynthesis—the former for structure and the latter for food—can be regarded as being formed by linking glucose molecules together into ribbons (cellulose) or bushes (starch). For

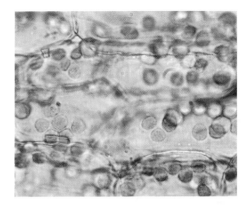

The small, round, green organelles in this section through the leaf of a snake plant (Sansevieria zeylanica) are the chloroplasts, the site of the photosynthetic apparatus of the plant.

When a leaf dies, the chlorophyll molecules decompose more rapidly than some of the other pigments; as a result, the colors of the latter—typically yellow, orange, and red—are no longer masked.

simplicity, we shall often write carbohydrates as $[CH_2O]$ and not display their structural complexity explicitly.

The overall reaction that carbon dioxide undergoes in the stroma of a chloroplast is

$$CO_2 + H_2O \longrightarrow [CH_2O] + O_2$$

At first sight, it looks as though the reaction involves stripping a carbon atom out of carbon dioxide, liberating an oxygen molecule, and linking the carbon atom to water to form the carbohydrate. However, although this interpretation was held for a long time, some circumstantial evidence noticed in the 1930s suggested that that is not in fact the overall scheme of the reaction. Certain bacteria (purple sulfur bacteria) thrive on the sulfur analog of water, hydrogen sulfide, H_2S, and a by-product of their photosynthetic activity is sulfur (S):

$$CO_2 + 2H_2S \longrightarrow [CH_2O] + H_2O + 2S$$

These deposits of sulfur, which have been mined in Canada, were formed by bacteria that used hydrogen sulfide, H_2S, in place of water, H_2O, as their source of hydrogen atoms and electrons. Unlike oxygen, sulfur is a solid and does not simply mingle with the atmosphere, but lies trapped beneath rock formations.

The sulfur, a solid, collects as globules in their cells rather than simply blowing away as a gas, as in the case of oxygen, and it is this accumulation that is responsible for the great deposits of elemental sulfur in various parts of the world. Quite plainly, the carbon atom has not surrendered the oxygen atoms that accompanied it into the cell: the sulfur comes from the hydrogen sulfide. The graduate student at Stanford University who noticed this analogy, C. B. van Niel, made the bold extrapolation that the pattern of photosynthetic activity shown by purple sulfur bacteria was common to aerobic photosynthesis, and therefore that *the oxygen came from the water:*

$$CO_2 + 2H_2O \longrightarrow [CH_2O] + H_2O + O_2$$

This conclusion was later confirmed using water molecules containing a radioactive isotope of oxygen, oxygen-18. When the course of the radioactivity was followed, it was found that the oxygen-18 that begins in the water ends up as a gas. We now see that the photosynthesis reaction involves stripping the hydrogen atoms off a water molecule and releasing oxygen. This component of the entire complex of reactions that account for photosynthesis is called the "water-splitting reaction." The oxygen we breathe, and which Faraday needed for his flame, was once a component of water, as we mentioned in Chapter 1, but now we can see that it was formed when the water was stripped of its hydrogen atoms. In photosynthesis, water is mined for its hydrogen, not for its oxygen.

The stumbling of organisms into discovering the water-splitting reaction led to an extraordinary outburst of evolutionary activity. Hitherto, anoxygenetic photosynthetic bacteria had clutched at organic acids and simple inorganic compounds for their hydrogen atoms (and, as we shall see, the electrons that are essential to photosynthesis). But, these compounds are rare, and so the niches that could support life were scarce. Then suddenly, about three billion years ago, an organism hit upon water as the source of hydrogen and electrons. That organism was the Columbus of nature, for its horizons were suddenly unlimited. No longer did it need to sail close to the shore of life where a scarce substrate was available: now life could inhabit every niche in the world.

There were two prices to pay for this adventurous opportunism. One was the onset of an extraordinary wave of global pollution, when the fecundity of the newly thriving novel organisms transformed the atmosphere by generating vast quantities of the deadly gas oxygen from the water that had been split. We can tell the moment of the Great Oxygenation, for the earth rusted, and many of our oxide ores date from that era. The other price concerned the organisms themselves, for the decomposition products of water—the intermediates produced on its progress toward becoming oxygen—are intensely dangerous and, if allowed to react with the other substances in the cell, could kill it.

REACTIONS THAT CAPTURE LIGHT

We can start to unravel the details of the reaction within the chloroplast by considering it to take place in two stages. One is the harvesting of the energy of the sun. The other is the deployment of that energy to convert carbon dioxide to carbohydrate. The site of the light-capturing reactions is one of the photosystems in the thylakoid membrane; the formation of carbohydrate itself takes place in the surrounding stroma.

Chlorophyll molecules are the antennae that respond to the incoming light. Each photosystem contains hundreds of chlorophyll molecules located

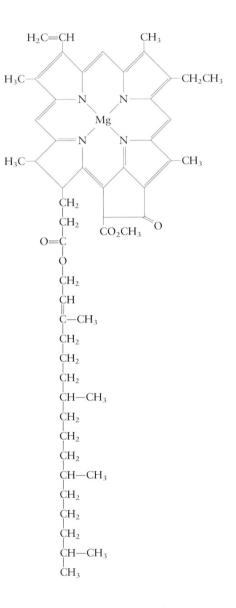

The chlorophyll a molecule, shown in the margin, consists of a "porphyrin" ring at the center of which is a single magnesium atom (Mg). The ring is the location of the light-capturing event, in which an electron is moved from one distribution around the ring to another of higher energy. The long hydrocarbon tail of the molecule does not affect the light-absorption characteristics of the molecule, but helps to anchor it in the hydrocarbon membranes within the chloroplast. Chlorophyll b differs only in the presence of a —CHO group in place of the —CH₃ at the top right of the molecule. The presence of this group modifies the light-absorbing properties of the porphyrin ring slightly.

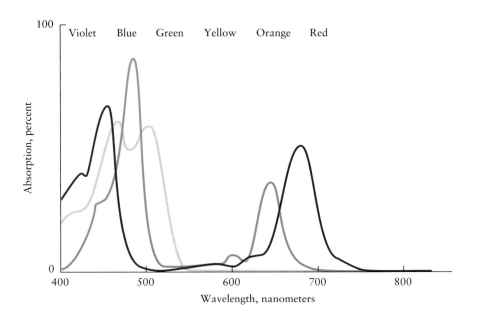

The absorption spectra of chlorophyll a *(red line) and chlorophyll* b *(blue line), and some of the carotenoids (yellow line) that also occur in the chloroplast.*

like an array of astronomical telescopes in the light-harvesting complex (LHC) of the photosystem. A chlorophyll molecule is ringlike, built around a single magnesium atom and having a long hydrocarbon tail. Its electrons are quite mobile, and one of them can be raised into an excited state by the absorption of a photon. A chlorophyll molecule can be excited by a photon of red light or a photon of blue light. The molecule does not absorb green light from sunlight, but reflects it back or allows it to pass through; this is the origin of the green color of the world's vegetation. There are in fact two important versions of chlorophyll, chlorophyll *a* and chlorophyll *b*, which differ in only one group that is attached to the ring and which have slightly different absorption characerics. The two types of molecules span a greater region of the incident light than either one alone, so they jointly harvest more energy. One route open to natural evolution (and to genetic engineering) is presumably to fill the gap in the absorption spectrum by developing new varieties of chlorophyll (the carotene molecules do that to some extent) and hence to harvest more energy from the sun. The aesthetic price is that vegetation would then be black, for all light would be absorbed.

Even at the instant of excitation the plant is confronted by its first difficulty, for an electronically excited chlorophyll molecule readily discards its energy as fluorescence. To avoid fluorescence, the photosystem adopts the strategy of funneling the excitation rapidly to a slightly modified chlorophyll molecule that has a slightly lower excitation energy than the antenna molecules. This molecule lies at the "reaction center" of the photosystem. Since the excitation energy of the molecules at the reaction center is slightly less than that of the antenna molecules, once the excess energy has been lost as heat, the excitation is trapped at the reaction center, like a ball in a billiard-table pocket. There are in fact two types of photosystem, called P1 and P2. The modified chlorophyll molecule is called P700 in Photosystem 1 and P680 in Photosystem 2, where P stands for pigment and the numbers 700 and 680 denote the wavelength of the light (in nanometers) that the molecule absorbs most strongly.

Now the excitation is at a location where it can be channeled appropriately. We shall consider Photosystem 2, but similar remarks apply to Photo-

Photosystem 2, the complex of molecules in the thylakoid membrane of chloroplasts that is responsible for trapping light and producing oxygen from water.

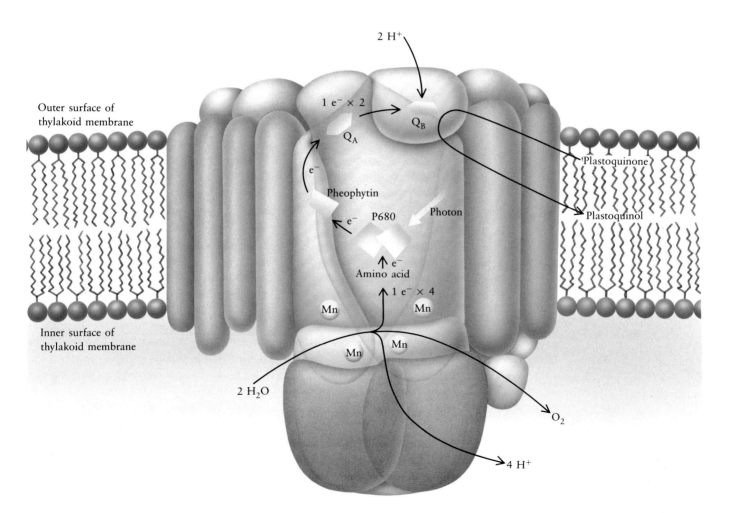

system 1, at least in the initial stages. The excitation is still liable to be lost uselessly as fluorescence, and so events must follow fast. To prevent fluorescence, the excitation energy must be converted into another form. To this end, a dramatic event takes place that will prove to be the cornerstone of photosynthesis and hence of life: an electron moves. The energy stored in the excitation of the P680 molecules is like a coiled spring that, suddenly released, ejects an electron to a neighboring molecule. The recipient molecule is called pheophytin; it resembles a chlorophyll molecule closely (for nature is economical as well as subtle), but lacks the latter's magnesium atom. Now the reaction center consists of a negatively charged pheophytin molecule (it is negatively charged on account of its additional electron) and a positively charged P680 chlorophyll molecule (it is positively charged on account of its loss of an electron). At this stage, the energy of the light has been transferred into the energy of spatial separation of positive and negative charges.

The risk of return of the electron to the P680 ion is still high, and if the electron were to snap back to its parent, the energy could still be lost as radiation or heat. The solution adopted by the photosystem is to separate the charges even further so that they no longer influence each other. At this stage, the reactions in the photosystem act like a ratchet, winching the electron up to the outer surface of the thylakoid membrane and lowering the positive charge down to the inner surface.

The migration of the electron to the outer surface is achieved in two stages. First, the electron is transferred from the pheophytin molecule to a nearby molecule Q_A (a plastoquinone), and the resulting ion immediately hands the electron, like a hot potato, to another plastoquinone molecule Q_B. The plastoquinone ion snips a hydrogen ion from the fluid in its environment, and then lies dormant until a second electron is passed to it by Q_A after a second photon-absorption event has occurred in the photosystem. When it snips a second hydrogen ion from its surroundings, the resulting plastoquinol molecule can break free from its location and migrate through the photosystem.

The plastoquinol molecule is a raft for the electrons that were released from the reaction center. It drifts out of Photosystem 2 and into the thylakoid membrane with its precious burden. Of this burden, more anon.

Back in the photosystem, we still have the P680 ion. This ion must also lose its charge and, in the process, achieve change. The first step is to fill the gap left by the departure of the electron when the ion was formed, for it is important for the reaction center to be ready to respond to the presence of another photon as soon as possible. This first step involves the transfer of an electron from a specific amino acid on a nearby protein molecule, and the second stage is the replacement of that electron by the ultimate source of

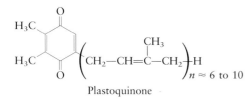

Plastoquinone

electrons in the plant—a water molecule. In the latter step, however, the system needs to remove no less than *four* electrons (e^-) from two water molecules to achieve the reaction

$$2H_2O - 4e^- \longrightarrow 4H^+ + O_2$$

This is the moment when water is split and oxygen is released; indeed, this reaction is the spring of our atmosphere.

As we have remarked, splitting water is a dangerous operation, for the intermediates can be lethal; there will be intermediates, because each photon absorption event generates a single P680 ion. Hence only one electron can be extracted from a water molecule in the wake of each photon, so splitting cannot occur with the single blow of a chemical ax. The cell has conquered this problem by lighting upon an enzyme system that contains the element manganese (Mn). As a consequence of the electronic structure of its atoms this element can lose a number of electrons, and in certain states it gives rise to relatively loose geometrical structures. The protein molecule that handles water splitting has an active site that consists of four manganese atoms at alternate corners of a cube; the other corners are occupied by four oxygen atoms. This little region, a chemical engine, is where the reaction occurs that sucks in water molecules and blows out oxygen molecules— ultimately, for us to breathe.

The sequence of events that gingerly takes two water molecules and strips them of their hydrogen atoms has been called the "water splitting clock," for it proceeds in five great ticks each time an electron is extracted from the active site of the enzyme in response to the formation of a P680 ion in the photosystem. Tick 1 sucks an electron from one of the manganese atoms; simultaneously a hydrogen ion, H^+, is released. The new, oxidized state of the manganese atom is reasonably stable and can survive long enough for the next demand by a newly created P680 ion. Tick 2, a second oxidation, removes a second electron from a manganese ion at the active site. The next tick of the clock is more complex: another electron is removed by a newly formed P680 ion, and a hydrogen ion is also lost; but as well as these losses taking place, two water molecules tumble in from the surrounding fluid and their oxygen atoms are used to form two new Mn—O—Mn links. Now the center consists of the same four manganese ions, but they form a cage with *six* oxygen atoms.

In the next tick of the clock, another electron is sucked out of the active site (it is thought that it comes from an amino acid that lies close to the manganese atom cage). That tick results in a highly stressed cage of manganese and oxygen atoms, and it is supposed that in the final tick the cage crumbles back into its original cubic form, squeezing out a molecule of

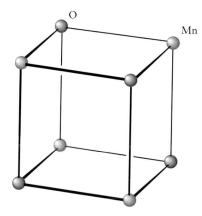

The active site of a photosystem, where water molecules are split, consists of this cube of alternating manganese (Mn) and oxygen (O) atoms. This regular structure is the heart of the photosynthesis apparatus: it is also the source of the majority of atmospheric oxygen.

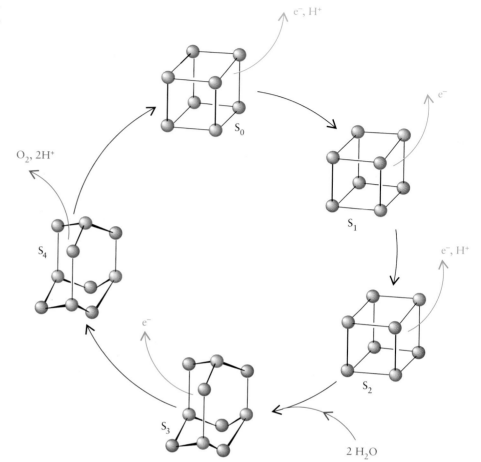

The manganese sites that are responsible for the water-splitting reaction, and the steps of the water-splitting clock.

e^-, H^+

e^-

S_0

S_1

$O_2, 2H^+$

e^-, H^+

S_4

S_2

e^-

$2 H_2O$

S_3

oxygen (and two more hydrogen ions) as it does so. The net effect of these ticks of the clock and the collapse of the cage is the loss of four electrons from two water molecules, the loss of their two oxygen atoms as an oxygen molecule, and the release of four hydrogen ions into the space within the thylakoid. The four electrons have effectively replaced the four that are riding through the membrane on two plastoquinol rafts. Now the charge separation is at its greatest, with the electrons on the rafts and the hydrogen ions snipped from water molecules in the interior of the cell.

One of the rafts and its two passengers floats through the membrane and meets another series of species. We shall not follow these complicated events in detail, except to say that at one of their encounters, the two electrons slip off the raft and are passed along a chain of molecules into Photosystem 1. The two hydrogen ions that were also passengers on the raft are tipped off into the interior of the thylakoid, so enhancing even further the acidity of the medium inside the membrane.

At this stage we can see the net effect of the absorption of four photons of light by Photosystem 2. Two water molecules have been split. An oxygen molecule has been formed. The acidity inside the thylakoid has been increased by the arrival of six hydrogen ions (four from the water splitting and two by transport on the raft from the stroma). Four electrons have been passed to Photosystem 1, and Photosystem 2 is back in its original state, ready to act again.

Meanwhile, the light-harvesting complex of Photosystem 1 has also captured a photon, and the excitation it causes has rattled from chlorophyll molecule to chlorophyll molecule until it is trapped at the P700 chlorophyll molecule at the reaction center of that photosystem. That molecule too must deploy its energy before losing it as a burst of fluorescent radiation. It passes on its hot potato by much the same mechanism as in Photosystem 2, by kicking out an electron. However, instead of the elaborate water-splitting reaction that restored the missing electron to the P680 ion, the P700 ion recovers the electron supplied by Photosystem 2. Now the overall net effect is seen as the splitting of water and the transfer of an electron to the pheophytin molecule in Photosystem 1.

That electron is not destined to float away on a plastoquinol raft through the pea-green sea. Instead, its task is to transfer to a particular molecule, the coenzyme nicotinamide adenine diphosphate (NADP), and to convert it into a different form of NADP, which is denoted NADPH. A coenzyme is a nonprotein molecule that assists an enzyme to carry out its task. Many vitamins are coenzymes or parts of coenzymes; nicotinamide, a part of the NADP molecule, is the vitamin niacin.

The only part of the tale still to be told is the role of the enhanced acidity within the thylakoid. It will be remembered that the water-splitting reaction strips four hydrogen ions off two water molecules inside the thylakoid; moreover, the formation of NADPH from NADP consumes hydrogen ions in the stroma, decreasing the acidity of the medium there. Overall there is a reservoir of hydrogen ions poised to surge out of the thylakoid like a waterfall; their concentration inside the thylakoid is about 10,000 times higher than that in the stroma. The outward flow of hydrogen ions occurs through small assemblies of proteins that span the thylakoid membrane; the unit acts like a chemical hydroelectric plant, and the energy released as the acidities equalize is used to build adenosine triphosphate (ATP) from adenosine diphosphate (ADP). As we have seen earlier, ATP is the molecule that acts as the immediate supply of energy in biological cells, and it gives up that energy when it reverts to ADP. Its presence in cells dictates the need for phosphorus in the diet, which is why the chemical industry is geared to send phosphates on their way through the biosystem as fertilizers. The rush of hydrogen ions stored in the thylakoid by the photon-absorption events is, in effect, used to recharge exhausted ADP molecules.

Before moving on, we should appreciate what the sequence of events has achieved so far. The energy-releasing ATP molecule has been recharged from ADP, so in the stroma there is a source of energy for enzymes to carry out their tasks. Moreover, through the action of Photosystem 1, there is also a ready supply of NADPH, the coenzyme that enzymes need to carry out their duties. The stroma is primed for action.

THE DARK REACTIONS

The stroma is poised for its specific action, the construction of carbohydrate from carbon dioxide. The carbon dioxide is there, the energy needed to link carbon atoms together is there, and the enzymes and their cofactors are also there and ready to carry out their tasks. Now we turn from the thylakoid to the stroma, to see the nature of the "dark reactions" that convert carbon dioxide to carbohydrate. The dark reactions do not occur in the dark (they are powered by incident light and its generation of ATP and NADH), but they do not require light directly.

The sequence of reactions that take place in the stroma was identified by Melvin Calvin of the University of California, Berkeley, and his colleagues in work that started in 1945 and culminated in the Nobel Prize in 1961. The first step in the "Calvin cycle" of reactions is one in which the carbon dioxide molecule containing the carbon atom that is destined to become organic is attached to a five-carbon molecule, ribulose 1,5-biphosphate (RuBP):

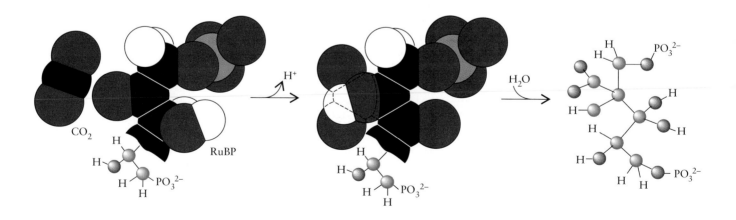

The enzyme that achieves this linking, which is called "rubisco" (more formally, ribulose 1,5-biphosphate carboxylase), is very abundant in chloroplasts, comprising about one-sixth of their total protein. Since chloroplasts are so abundant, rubisco is probably the most abundant protein in the world.

Once the six-carbon molecule is formed, it is promptly snipped by an enzyme into two three-carbon ions:

Melvin Calvin (b. 1911).

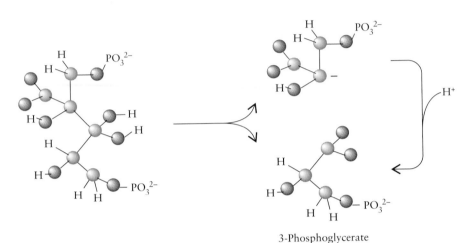

3-Phosphoglycerate

Now a complex series of rearrangements ensues, summarized in the illustration on page 225. In broad outline, for every six CO_2 molecules that are brought into the cycle by attaching to six RuBP molecules with the production of twelve three-carbon ions, ten of the latter are used to reform six RuBP molecules so that the cycle can continue. The remaining two three-carbon ions are linked together to form a six-carbon molecule, glucose, which is the net profit of the transaction. Carbon dioxide has become carbohydrate, and the cycle is driven by the ATP and NADH formed in the thylakoid.

It turns out, however, that nature appears to have made a design fault in the Calvin cycle in an overhasty adoption of rubisco as the enzyme. The fault arises because rubisco was developed early in the history of photosynthesis, when carbon dioxide was abundant in the atmosphere and there was very little oxygen. The particular fault that has become apparent is that rubisco also catalyzes a reaction that competes with the linking of carbon dioxide to RuBP. In the competing reaction, oxygen, not carbon dioxide, is added to RuBP to form one three-carbon molecule and one two-carbon molecule:

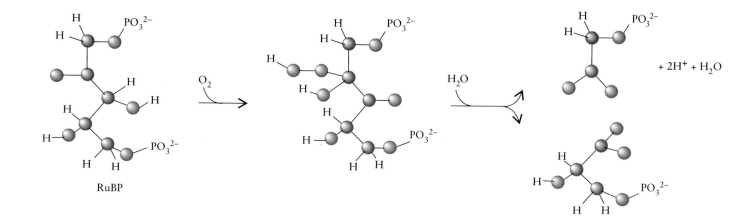

RuBP

This reaction is particularly important now that oxygen is so abundant. A series of reactions seeks to repair the damage by managing to convert two of the two-carbon molecules into a three-carbon molecule plus one one-carbon molecule; but that one-carbon molecule is carbon dioxide! Thus, rubisco had defeated the whole elaborate network of reactions by taking oxygen and converting it into carbon dioxide—the very opposite of what the chloroplast aims to achieve. This unwelcome path, the analog of respiration in animals (but without the beneficial production of ATP that occurs in animal respiration), is called "photorespiration." Under normal conditions, as much as half the carbon fixed by photosynthesis may be burned away to carbon dioxide by photorespiration. The problem is particularly severe in tropical plants because the rate of photorespiration increases more rapidly with temperature than does that of the carbon-fixing reaction, and so in hot climates the net rate of photosynthesis may be low.

This serious fault can perhaps be repaired by genetic engineering. If photorespiration were to be successfully suppressed, the engineered plant would be able to process carbon dioxide significantly more efficiently, thus resulting in greater crop yields even in temperate climates. Such studies are currently in progress. Nature has itself recognized the fault and has experimented with the evolution of an alternative pathway, known as the C_4 pathway, or the Hatch-Slack pathway, after the Australian plant physiologist M. D. Hatch and C. R. Slack, who established it. In C_4 plants, the carbon dioxide molecule is linked to a three-carbon molecule to give a four-

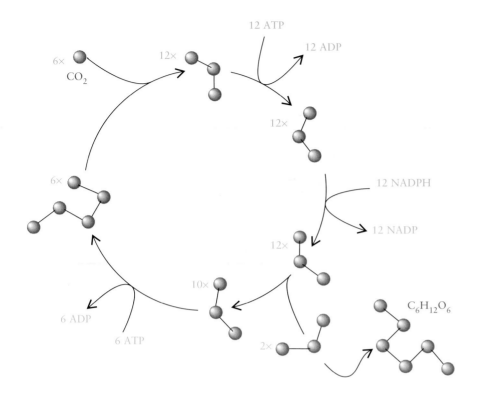

12 ATP

12 ADP

6× CO₂

12×

12×

12 NADPH

12 NADP

12×

10×

6 ADP

6 ATP

2×

C₆H₁₂O₆

6×

The Calvin cycle. On each circuit, which is driven by the ATP and NADPH molecules produced by the action of light, one CO₂ molecule is taken in. The results of six successive cycles are shown here. The six cycles take in six CO₂ molecules and combine with six molecules of RuBP (a C₅ compound) to produce 12 molecules of a C₃ compound (3-phosphoglycerate). These 12 molecules are converted first into 12 molecules of diphosphoglycerate and then into 12 molecules of another C₃ compound (glyceraldehyde 3-phosphate), and 10 of these molecules combine to form six RuBP molecules, resulting in a net gain of two C₃ molecules, which combine to form one C₆ sugar molecule.

carbon intermediate. This four-carbon intermediate, which is formed outside the chloroplast in the mesophyll of the leaf, breaks down into carbon dioxide in the region of the chloroplast, and that carbon dioxide enters the Calvin cycle. The C_4 cycle makes heavier demands on the energy production of the cells involved, because whereas C_3 plants need three ATP molecules to fix one carbon dioxide molecule, a C_4 plant needs four. However, the advantage of the C_4 pathway is that photorespiration in such plants is nearly completely absent. This is because the C_4 pathway leads to a very high local concentrations of carbon dioxide in the region where the Calvin cycle is taking place, so that the carbon-fixing reaction becomes dominant. In effect, the C_4 pathway recreates a microenvironment within the leaf that is similar to the carbon-dioxide-rich environment in which rubisco originally developed. Overall, therefore, the process may be more efficient even though it makes heavier demands of ATP.

The provision of carbon dioxide by this mechanism also helps the plant to make more efficient use of sunlight. In tropical regions, the supply of carbon dioxide limits the rate at which photosynthesis can occur, but the C_4 mechanism helps to pump the meager supply to the exact location it is needed. The contribution of the C_4 pathway can be quite substantial, for the net photosynthetic rates of C_4 grasses such as corn and sugarcane can exceed by a factor of two or three the rates of C_3 grasses such as rye, oats, and rice.

The role of the C_4 pathway is apparent in lawns, which in cooler climates consist of C_3 grasses, such as creeping bent. However, in the heat of summer, the C_4 gasses like crabgrass with their broader, paler leaves, may overwhelm the darker, finer C_3 grasses as the more efficient C_4 pathway delivers carbon dioxide to the Calvin cycle.

In this account we have merely touched on the complex events that take place in an illuminated leaf. But all those events—in particular the pregnant transition that carbon dioxide makes when it enters the biosphere—are strings of chemical reactions. In his demonstrations with candles, Faraday was making use of the fruits of photosynthesis, both in the waxes of the candle and the oxygen that was used to sustain the flame. He was also participating in the great cycle of nature, for in his conjuring with flames, he was liberating the dormant carbon atoms of the wax and enabling their return to life.

EPILOGUE

Faraday lit his candle in the gloom of mid-nineteenth-century London, and opened the eyes of his young audience to the current understanding of the composition and transformation of matter. The contents of his lectures epitomized the contemporary status of chemistry: a subject that had begun to accumulate concepts yet was ignorant of many of the concepts that we now know to be pivotal. The atom was known, the electron not; mechanics was known, but quantum mechanics not. Then the concern of chemistry was to accumulate a familiarity with the transformations that substances could undergo, the explanation of these changes lay decades in the future.

Now, when we light a candle, we can imagine the dance of the atoms, the shift of electrons, and the formulated yet still only marginally comprehensible role of quantum mechanics. In the century and a half since Faraday, we can understand the inner, if not the innermost, workings of matter, and have shifted the focus of our attention from the transformation of substance to the understanding of that transformation in terms of the atomic and the subatomic.

The understanding that chemists have achieved has given them unparalleled control over matter. Now they can construct intricate molecules of hundreds of atoms linked in a filigree of bonds. Now they can design a catalyst, and achieve speedily and economically a transformation that might lead to a whole new type of material, such as a polymer or a pharmaceutical. Now, too, they can peer into the workings of living cells and see, not only the mechanism of photosynthesis, but many of the processes that photosynthesis enables and which, if they did not occur, would leave life crippled.

But what of the future? What will a chemist, reflecting on a candle yet another century and a half into the future, think it appropriate to tell an audience? Where will chemistry have arrived after one more great tick of the clock? Now we are as blind as Faraday would have been had he considered this question in his time. However, although it may be tempting fate to say so, it may be that chemistry has reached a plateau on its conceptual landscape, and that extrapolation from today is more likely to be successful than it would have been in Faraday's day, before the nuts and bolts of the subject and their mechanics had been discovered. If chemistry now has all its hardware—needing nothing beyond atoms and electrons—and if it has its rules of the game—essentially quantum mechanics—then although there may be some technical surprises still to come (and countless major achievements), peering into the future through a speculating telescope may be dangerous but not ludicrous.

A century and a half hence, with Faraday now an even more dimly distant star, you will probably read of considerable achievements but not of revolution. Almost certainly the paradigm of discussion and the means of manipulation and discovery will be the computer. A chemist will not need to labor at the bench to achieve an intricate structure, but will design it on a screen, and leave the computer and its automatic synthesizers to assemble the molecule. What molecules to make will also be taken largely out of a chemist's hands, for compounds will be screened for pharmacological activity, color, texture, tensile strength, or whatever property is intended by the computer itself, and only those molecules that are appropriate, or nearly appropriate, will be made. Molecules that are too intricate to make in this way will be built, as some are already, by drawing on the stable of genetically-engineered bacteria that will act as our future flasks. Almost certainly, many chemical reactions will be carried out in living cells, with the human (computer-directed) chemist burnishing the final details of the compound. The distinction between inorganic and organic will finally have disappeared, and chemists will use bacteria to generate intricate arrays of atoms other than carbon, perhaps ending the synthesis by burning away the organic from the biological product, leaving the web of inorganic atoms.

The understanding of chemical reactions will also have reached new levels of detail, but the understanding will be recognizably an extension of what is already known. Already, at resolutions of 1 femtosecond (10^{-15} seconds), almost all nuclear motion is frozen, and chemistry has ceased. We shall soon be able to visualize the individual events taking place frame-by-frame every femtosecond, and understand the role of the solvent as well as the reacting solute. Such information will also be modeled by solving the Schrödinger equation, and displaying the results graphically. Then, perhaps, it will be hard to distinguish whether we are looking at the experiment or the simulation, so accurate will the calculations be as they draw on ever greater, globally networked computing power.

More elusive now, but perhaps yielding to scrutiny then, will be our ability to pursue the motion of individual electrons—to observe (or calculate, then visualize) electron distributions welling forward or drawing back as reactants approach through a solution, and then to observe the electronic upheaval that marks the reaction itself. When we can do all that, then we shall truly understand atoms, electrons, and change.

FURTHER READING

CHAPTER 1

Faraday's lectures were originally published as

M. Faraday, *A course of six lectures on the chemical history of the candle*, Royal Institution of Great Britain, London (1861).

They are currently available, with additional experiments, as

Faraday's chemical history of a candle, Chicago Review Press, Chicago (1988).

CHAPTER 2

The concepts of quantum mechanics that are relevant to chemistry are set out in a pictorial, nonmathematical way in

P. W. Atkins, *Quanta: A handbook of concepts* (2nd edition), Oxford University Press, Oxford (1991).

CHAPTER 3

For an introduction to the range of chemical reactions, see

P. W. Atkins, *General chemistry*, Scientific American Books, New York (1989).

CHAPTER 4

For a nonmathematical account of entropy and the Second Law of thermodynamics, see

P. W. Atkins, *The second law*, Scientific American Library, W. H. Freeman, New York (1984).

For a more quantitative treatment, see

P. W. Atkins, *Physical chemistry* (4th edition), Oxford Univerity Press, Oxford, and W. H. Freeman, New York (1990).

CHAPTER 5

The temperature dependence of chemical reactions is described in

K. J. Laidler, *Chemical kinetics* (3rd edition), Harper and Row, New York (1987).

The mechanism of reactions in terms of collision between molecules is reviewed in

J. N. Murrell and S. D. Bosanac, *Introduction to the theory of atomic and molecular collisions*, Wiley, Chichester (1989).

L. Salem, *Electrons in chemical reactions*, Wiley, New York (1982).

The quantum mechanical effects of tunneling are surveyed in

R. P. Bell, *The tunnel effect in chemistry*, Chapman and Hall, London (1980).

An introductory article describing femtosecond observations is

A. H. Zewail, The birth of molecules, *Scientific American*, **263**(6): 40 (1990).

CHAPTER 6

For an account of chemical kinetics, see the book by Laidler cited above.

The review article on which the account of oscillating reactions is based is

R. J. Field and F. W. Schneider, Oscillating chemical reactions and nonlinear dynamics, *J. Chem. Educ.* **66**: 195 (1989).

The same issue of the journal contains several articles in which oscillations and chaos are discussed.

For a survey of chemical oscillation and chaos, see

P. Gray and S. K. Scott, *Chemical oscillations and instabilities*, Oxford University Press, Oxford (1990).

CHAPTER 7

Two excellent guides, the first introductory and the second authoritative, to the mechanism of organic reactions are

P. Sykes, *A guidebook to mechanism in organic chemistry* (6th edition), Longman, London (1986).

T. H. Lowry and K. S. Richardson, *Mechanism and theory in organic chemistry* (3rd edition), Harper and Row, New York (1987).

Although this book has not dealt with the mechanism of reactions of inorganic compounds (the subject is in a more primitive state than that of organic reactions), an introduction will be found in

F. Basolo and R. G. Pearson, *Mechanisms of inorganic reactions,* Wiley, New York (1967).

K. Katakis and G. Gordon, *Inorganic reaction mechanisms,* Wiley, New York (1987).

CHAPTER 8

The books dealing with organic reaction mechanisms also describe the conservation of orbital symmetry, but see also

R. B. Woodward and R. Hoffmann, *The conservation of orbital symmetry,* Academic Press, New York (1970).

T. L. Gilchrist and R. C. Storr, *Organic reactions and orbital symmetry* (2nd edition), Cambridge University Press, Cambridge (1979).

J. Simons, *Energetic principles of chemical reactions,* Jones and Bartlett, Boston (1983).

CHAPTER 9

For an excellent account of photosynthesis, see

L. Stryer, *Biochemistry* (3rd edition), W. H. Freeman, New York (1988).

An introductory article on the water-splitting reaction (with further references) is that by

Govindjee and W. J. Coleman, How plants make oxygen, *Scientific American,* **262**(2): 42 (1990).

For the chemical generation of light, see

R. P. Wayne, *Principles and applications of photochemistry,* Oxford University Press, Oxford (1988).

SOURCES OF ILLUSTRATIONS

Illustrations by Tomo Narashima and Fineline.

FACING PAGE 1   J. B. Diederich/ Woodfin Camp & Assoc.

PAGE 2   The Director of the Royal Institution

PAGE 3   The Science Museum, London

PAGE 4   Chip Clark

PAGE 5   Nishiinoue/Orion Press/ Sipa Press

PAGE 6   Randall M. Feenstra, IBM Thomas J. Watson Research Center, Yorktown Heights, N.Y.

PAGE 8   Image of the structure of L. casei thymidylate synthase, Hardy, et al. Science, Vol. 235, pp. 448–455, 1987 in the lab of Robert M. Stroud; graphics by Julie Newdoll using MIDAS+ software from the UCSF Computer Graphics Laboratory.

PAGE 9   Chip Clark

PAGE 10   The Science Museum, London

PAGE 12   Camille Vickers, Consolidated Natural Gas

PAGE 13   Runk/Schoenberger/Grant Heilman Photography

PAGE 14   Michelle Vignes

PAGE 15   IBM Research Division/ Almaden Research Center

PAGE 16   David Malin, Anglo-Australian Observatory

PAGE 25   Chip Clark

PAGE 26   Chip Clark

PAGE 32   University Archivist, The Bancroft Library, U. of California, Berkeley

PAGE 38   Pacific Gas and Electric

PAGE 42   Chris Pellant

PAGE 45   Peter Arnold, Inc.

PAGE 47   Kungliga Vetenskap-sakademien

PAGE 48   Ken Karp

PAGE 49   (left) Det Kongelige Bibiotek

PAGE 49   (right) University Chemistry Laboratory, University of Cambridge

PAGE 51   Aalborg Portland Beton-forskningslaboratorium, Karlslunde

PAGE 52   Chip Clark

PAGE 54   Chip Clark

PAGE 56   Peter Kresan

PAGE 63   (left) J. B. Diederich/ Woodfin Camp & Assoc.

PAGE 63   (right) Woods Hole Oceanographic Institution

PAGE 64   (left) Ken Karp

PAGE 64   (right) Joe Viesti/Viesti & Assoc.

PAGE 67   Arianespace, Inc.

PAGE 70   Chip Clark

PAGE 74   ONERA

PAGE 76   Yale University Archives

PAGE 77   The Metropolitan Museum of Art, Ford Motor Company Collection, Gift of the Ford Motor Company and John C. Waddell, 1987. (1987.1100.32)

PAGE 79   (top) Chip Clark

PAGE 79   (bottom) Manley-Prim Photography

PAGE 85   Chevron Corporation

PAGE 88   Ken Karp

PAGE 89   The Director of the Royal Institution

PAGE 90   The Director of the Royal Institution

PAGE 91   Burndy Library

PAGE 92   Ken Karp

PAGE 94   Alexander McPherson

PAGE 100   Chip Clark

PAGE 103   Kenneth Lorenzen

PAGE 108   Deutsches Museum, München

PAGE 111   George Hall/Woodfin Camp & Assoc.

PAGE 112   Chip Clark

PAGE 113   Catalytic Systems Division, Johnson Matthey

PAGE 125   (top) M. Karplus, R. N. Porter, and R. D. Sharma, *J. Chem. Phys.* **43**: 3258 (1965).

PAGE 125   (bottom) P. Brumer and M. Karplus, *Faraday Disc. Chem. Soc.* **55**: 80 (1973).

PAGE 126   Ahmed H. Zewail, California Institute of Technology

PAGE 127   Ahmed H. Zewail, *Science* **242**: 1645 (1988). © 1988 by the AAAS.

PAGE 128   Larry Brownstein/Rainbow

PAGE 131   Ken Karp

PAGE 135   (left) Bruno Barbey/Magnum

PAGE 135   (right) Chip Clark

PAGE 140   Dr. Syun Akasofu, Geophysical Institute, University of Alaska

PAGE 144   Fritz Goro

PAGE 154   (top) J. C. Roux, R. Simoyi, and H. Swinney, *Physica* 8D: 257 (1983).

PAGE 154   (bottom) Ken Karp

PAGE 155   Ken Lucas/Planet Earth Pictures

PAGE 156   J. D. Murray, *Mathematical Biology,* Springer-Verlag, 1989.

PAGE 158   Gary Braash/Woodfin Camp & Assoc.

PAGE 166   Ken Karp

PAGE 173   Mula and Haramaty/Phototake

PAGE 177   (top) IBM Research Division/Almaden Research Center

PAGE 180   Kenneth Lorenzen

PAGE 183   (left) Harvard University Archives

PAGE 183   (right) Roald Hoffmann

PAGE 202   Ric Ergenbright photography

PAGE 205   Runk/Schoenberger/Grant Heilman Photography

PAGE 206   (top) Steve Solum/Bruce Coleman Inc.

PAGE 206   (bottom) Gregory K. Scott/Photo Researchers

PAGE 207   Chip Clark

PAGE 211   Ray F. Evert

PAGE 212   Grant Heilman/Grant Heilman Photography

PAGE 213   (top) Runk/Schoenberger/Grant Heilman Photography

PAGE 213   (bottom) Gary Meszaros

PAGE 214   Chevron Corporation

PAGE 217   Govindjee and William J. Coleman, How plants make oxygen, *Scientific American* **262**(2): 50–58 (February 1990). © Scientific American, 1990. All rights reserved.

PAGE 223   University Archivist, The Bancroft Library, U. of California, Berkeley

INDEX